T0306030

Enterprise Project Management

Enterprise Project Management: A Comprehensive Guide to Successful Management by Projects covers the essential and fundamental topics of Enterprise Project Management and Management of Change by projects. It is written for portfolio, program, and project managers, members of the project community, upper- and middle-level management, functional and operational managers, and all who desire to acquire an understanding of effective change by project management. The book covers in-depth the following important aspects of Enterprise Project Management:

- Achieving organizational goals
- Management of programs
- Benefits realization management
- Stakeholder management and engagement
- Project Portfolio Management (PPM) and the Project Management Office (PMO)

The book explains how enterprises can consistently succeed in managing projects by aligning them with Business Goals and clearly defining what needs to be achieved. It shows how to ensure that Enterprise Project Management is fully deployed, and that project management concepts, methods, and techniques are available and utilized to deliver business value and realize benefits.

The book helps managers to answer the question, "Are we doing the right projects?" by covering how PPM can ensure project alignment with strategic or operational goals and the efficient use of scarce resources and funding to achieve Objectives and Goals. It also helps managers to answer the question, "Are we doing projects right?" by explaining the critical role of a PMO, which supports excellence in project management by enhancing the proficiencies of Project Managers and providing the foundational tools and techniques for project success.

Claude H. Maley is managing director of MIT Consultants, a consultancy and education practice servicing international clients in management of change, and chairman of a business solutions company. He started his career as a systems engineer with IBM, after reading estate management and building construction at the London School of Building, and has held various management positions for international organizations and companies.

Enterprise Project Management
A Comprehensive Guide to Successful Management by Projects

Claude H. Maley

CRC Press
Taylor & Francis Group
Boca Raton London New York

CRC Press is an imprint of the
Taylor & Francis Group, an **informa** business
AN AUERBACH BOOK

First edition published 2024
by CRC Press
2385 Executive Center Drive, Suite 320, Boca Raton, FL 33431

and by CRC Press
4 Park Square, Milton Park, Abingdon, Oxon, OX14 4RN

CRC Press is an imprint of Taylor & Francis Group, LLC

© 2024 Taylor & Francis Group, LLC

ISBN: 978-1-032-54373-4 (hbk)
ISBN: 978-1-032-45582-2 (pbk)
ISBN: 978-1-003-42456-7 (ebk)

DOI: 10.1201/9781003424567

Typeset in Typeset in Garamond
by SPi Technologies India Pvt Ltd (Straive)

Contents

Introduction: Introduction to Enterprise Project Management

Change is constant as socioeconomic, geopolitical, societal, environmental, and technological forces are not static. Change in any of these areas does not pursue a linear trajectory, and the combination of these forces imposes enterprises and organizations, be they private or public, to continuously evolve to maintain and sustain their operation and to compete and grow.

The dynamics of ever-growing multinational and local demands additionally oblige organizations to be nimble and flexible and manage change effectively to address these evolutionary drives. Enterprises must extract themselves from the pseudo-comfort that holds the belief that they can control change, and eventually their own destiny.

The conclusion is simple:

Change is here to stay and is the only constant in an environment of change.

Enterprises cannot expect to continue to perform and operate successfully relying only on their current organizational structure, people, and processes as these are not isolated from events, be they internal or external. Organizations must therefore have a mentality and approach to manage the internal changes they require and desire to accomplish and to respond to external changes that are forced upon it.

The ability to achieve organizational goals and to manage strategic and tactical operational change initiatives effectively and successfully is one of the most important competencies that organizations expect from their executives, managers, and employees.

DOI: 10.1201/9781003424567-1

The most acknowledged instrument of change for organizations is Management by Programs and Projects and its application in Enterprise Project Management.

Enterprise Project Management is a holistic ensemble. It covers organizational management, operational structure, extensive communication, and rational and coherent decision-making. It is applied in initiating, performing, and managing transformational and tactical change programs/projects integrally, as it adapts to the organization's evolving operational environment. Enterprise Project Management is prevalent in the organization's pursuit of its Strategic and Operational Objectives and in coordinating funding and resources in a direct relationship to its Vision, Mission, Strategy, and Goals. Enterprise Project Management provides an effective and powerful platform for the organization's collective efforts.

Enterprises, small-medium companies, governments, and nonprofit organizations have recognized that to be successful, they need to be conversant with and use project management methods and techniques. This is the focus that organizational management must pursue as it combines proactivity and reactivity to address its challenges for both sustaining and growing its operation.

The discipline, knowledge, application, and deployment of project management across the organization must take center stage and be supported by all levels of management for both strategic and tactical change. Placing a strong emphasis on Enterprise Project Management as the key enabler of successful change will enable organizations to increase their ability to meet their goals and fulfil their Objectives.

Organizations must develop and enhance the skills and discipline of managers and project professionals by learning, comprehending, adapting, and deploying a holistic framework to perform programs and projects so as to realize the desired results. Such a framework places a great emphasis on the organization's capability to institute, support, develop, and maintain a project management community that is highly competent and skilled.

Successful organizations recognize that experienced and highly skilled Project Managers are essential to project success. They also know that an effective Project Manager needs to be business orientated and understand strategic, financial, organizational, and commercial issues and challenges. Above all, they seek those Project Managers who possess strong communication and interpersonal skills. Companies that outclass in project management will ensure the retention of their Project Managers by providing them with career opportunities, training, and mentoring.

Furthermore, Project Managers require effective people skills to manage stakeholder expectations and strong relationship and communication abilities to address and resolve conflicts in accomplishing the project's Objectives. Those Project Managers who know how to influence people, can make things happen, ensure the co-operation of others, and maintain efficient and harmonious working relationships.

Enterprises that consistently succeed in managing projects align them with Business Goals and clearly define what needs to be achieved. They ensure that Enterprise Project Management is fully deployed and that project management concepts, methods, and techniques are available and utilized to deliver business value and realize benefits.

Projects compete for limited funding and resources. To answer the question, "Are we Doing the Right Projects?" the organization must institute a Project Portfolio Management system to ensure project alignment to the strategic or operational goals and efficiently employ scarce resources and funding to achieve Objectives and Goals.

To answer the question, "Are we Doing the Projects Right?" the organization must institute a Project Management Office. This structure will support the project management community to deliver excellence in projects by enhancing the proficiencies of Project Managers and providing the foundational tools and techniques for project success.

This book covers the essential and fundamental topics of Enterprise Project Management and Management of Change By Projects. It is addressed to program and Project Managers, members of the project community, upper- and middle-level management, functional and operational managers, and all who desire to acquire an understanding of effective change by Project Management.

The book consists of five chapters that combine to guide the reader's comprehension and understanding of Enterprise Project Management and the organizational and communication themes, methods, and techniques to be employed for success in managing projects:

Chapter 1 – Achieving Organizational Goals
Chapter 2 – Management of Programs
Chapter 3 – Benefits Realization Management
Chapter 4 – Stakeholder Management and Engagement
Chapter 5 – Project Portfolio Management (PPM) and the PMO

All chapters are cross-related to each other, allowing for a comprehensive understanding of Enterprise Project Management. Each chapter may be read independently, since all chapters cover key topics that overlap with each other to provide continuity. However, the author suggests that Chapter 1 be read first to gain a global perspective of the essence of Enterprise Project Management.

Claude H. Maley
January 2023

Chapter 1

Achieving Organizational Goals

1.1 Chapter Overview

This chapter lays the basis for Achieving Organization Goals, which cannot be accomplished without a comprehensive understanding of the strategic and operational initiatives undertaken and launched to successfully manage changes to the organization's operation.

Programs and projects are the result of selected initiatives and will create change. The launch decisions of these initiatives are made both at the strategic level of management and at the operational units. The organization must be geared to implement and deploy a Management of Change structure and instill a programs and projects culture across all operating units.

This chapter addresses the transformational and tactical change approaches that an organization must follow. Topics related to Initiatives and Projects, Business Benefits Management, Stakeholder Management and Project Portfolio Management, and the PMO are discussed. Each of these topics is further detailed in Chapters 2 to 5.

The reader is invited to reference the Project Management Institute's "Standard for Organizational Project Management (2018)", which presents a framework to align project, program, and portfolio management practices with Organizational Strategy and Objectives.

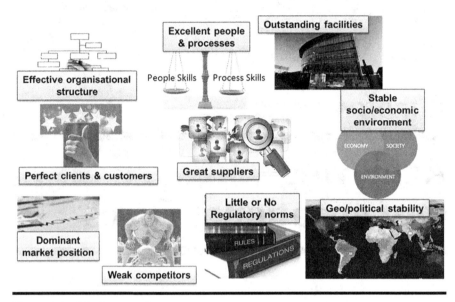

Figure 1.1 Ideal enterprise.

1.2 The Changing Face of Business

Upon the creation of an organization, be it private or public, and its subsequent operation, there are several aspects that need to function perfectly (Figure 1.1):

By starting with a solid organizational structure that is well established, and well distributed. Excellently trained staff with an elevated level of competence. Business processes that are perfectly smooth. The organization's facilities are of a high standard. The client base is extremely satisfied with the products and services offering. Suppliers, providers, and contractors are extremely proficient, and all deliveries are on time with high quality. The organization has a dominant market position and competition is weak. There are no impeding regulatory norms. The organization bathes in a socio-economic, and geopolitical stabilities.

This ideal picture is, however, far from the reality. The dynamics within the organization, and those external to it, will constantly evolve as change occurs. And all the changes would affect the organization's ability to perform to meet its desired Objectives.

Change is a constant, and for companies to achieve their organizational goals, a comprehensive understanding of the mechanics of the dynamics, and how to respond to these, must be incorporated at all levels of the organization.

1.2.1 The Reality of Business Change

In 2018, the Business Reality Check, developed by the Economist Intelligence Unit explored core business challenges that are critical to the successful execution of an organization's Mission, strategy, goals, and operations. Ten major challenges, or themes, were identified following a review of business, industry, and academic literature and interviews with senior executives.

Of these ten challenges, this chapter will revolve around the most salient and pertinent ones in Enterprise Project Management, such as:

- Managing through a shifting macroeconomic and political environment
- Managing operational complexity
- Talent management
- Meeting stakeholder expectations
- Managing shifting customer demands
- Managing the pace of technological change

The other themes are:

- Managing the supply chain
- Managing cross-border regulations
- Managing cybersecurity threats
- Using data effectively

There are multiple other challenges that organizations face depending on the nature of their operation, be it in private or public industry, and their geography.

However, all the above will force organizational change and demand that actions be planned and performed which are required to amend and evolve major components of the organization's operating environment, internal and external processes, physical and digital infrastructures, and most of all, culture with regard to changing environments and skills, competencies, and retention of employees.

1.2.2 Organization's Ability to Meet Market Requirements

The major challenge for any organization is to sustain and maintain its operation and pursue a constant need for continuous improvement and innovation. The organization will strive to balance its strategic direction with its operational performance while addressing situations that require it to be reactive or proactive.

Major internal forces are at play simultaneously, all contributing to contest the level of performance the organization aims to achieve (Figure 1.2).

The core forces requiring particular attention as to their performance are:

- Analyzing and reviewing the market, competition, services, and products on offer
- Delivering end-to-end solutions within time-to-market constraints
- Transforming key processes to strategic capabilities
- Redefining organizational structures and empowering people
- Facilitating enabling technology

Figure 1.2   Sustaining business operations.

1.2.2.1 Making Strategic Investments

All the above forces will experience internal changes because of their inherent characteristics, and the organization will be summoned to be proactive or else face a situation that calls upon it to be reactive.

The organization must also integrate the results of several external influences, such as evolving competition, financial constraints, evolving marketplace, temporal constraints, socioeconomic and geopolitical disturbances, impacts of political and regulatory changes, and globalization.

Operating in a stable environment is somewhat of a nonsense.

Change is the only certainty in a world of uncertainty.

1.2.3   The Organization's Strategic Intent

The core intents of the organization are meeting its Business Goals as it manages its performance, in a constant state of change, in an iterative and incremental manner, and in concert with continuous improvement, Business Process Reengineering, and facilitating enabling technology. Another core intent is to have a proactive rather than reactive operational approach so as to create value.

The organization will focus on two paths simultaneously: strategic transformation and tactical operation (Figure 1.3).

1.2.3.1 Strategic Transformational

The goals of strategic transformation are to evolve the organization in the medium and long term by adapting and progressing the organizational structure to meet anticipated

Figure 1.3   Value creation.

future demands; pursuing and stimulating growth in current products and services and developing new offerings that are in line with the strategic transformation goals.

Both the strategic transformational and Tactical Operational paths, especially continuous improvement for the latter, will form the organization's Change Management focus and actions. These actions will be initiatives that revolve around the following major areas:

- Performance to an elevated standard by focusing on core competencies; instituting and sustaining a learning organization; and empowering individuals and promoting self-directed work teams
- Operating to ambitious standards by sustained market and customer service; continuous improvement; total quality management (TQM); and reengineering/business process redesign and establishing strategic alliances

1.2.3.2 Tactical Operational

The performance goals of the organization are to sustain its continuous operation, while maintaining a competitive advantage with its products and services offering.

1.2.4 "Are We Doing It Right?"

Effective Strategic Transformational and Tactical Operational Change Management will constantly interrogate the organization with two simple questions – Are we Doing the Right Thing? Are we Doing the Thing Right? (Figure 1.4)

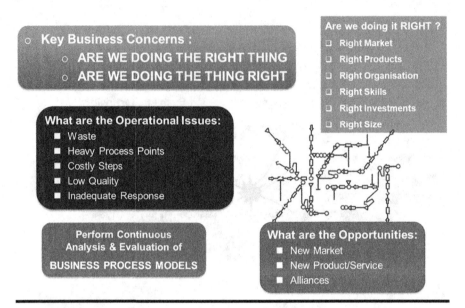

Figure 1.4 Doing it right.

Are we Doing the Right Thing will cover the right market, the right products, the right organization, the right skills, the right investments, the right size, and other areas specific to the operating environment of the organization.

Tactical Operational analysis will address issues covering – waste; heavy process points; costly steps; low quality; inadequate response and other areas specific to the operating environment of the organization. Operational performance will be a continuous daily, weekly, and monthly evaluation by measuring the status and progress of the operational units.

Strategic transformational analysis will focus on the continuous analysis and evaluation of business process models and the assessment of opportunities such as new market needs and new products/services to offer and alliances.

Strategic Transformational and Tactical Operational Change Management can only be successful once organizational initiatives have been identified and the corresponding programs and projects that enable the change to happen are launched.

1.2.5 Change as an Opportunity

Change projects in the Tactical Operational arena will focus on how to sustain and rationalize current operations by enabling focused responses to remove faults and obstacles, and by engaging in continuous improvement and TQM change initiatives (Figure 1.5).

Strategic transformation will focus on change initiatives that will enable innovation to gain competitive advantage and transformational redesign for performance improvement across all organizational units.

If you say "no" to change, change will say "no" to you.

Figure 1.5   Degree of change.

1.2.6 Strategy as a Method for Leveraging Change

Change, be it from internal or external forces, will affect different areas of the organization, and these areas will be interdependent and must be coordinated in unison for the organizational goals to be achieved (Figure 1.6).

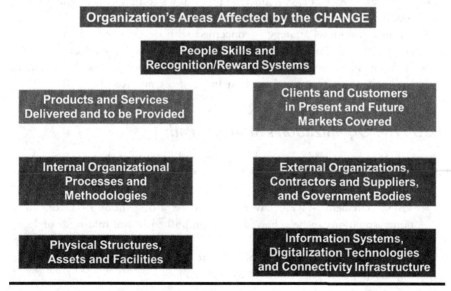

Figure 1.6   Areas affected by change.

Organizational goals are classified as follows: official, operative, operational, and departmental/unit. These may have different terminologies across different organizations.

- Official goals convey the values of the organization, through Vision and Mission statements
- Operative goals are indicators of the tangible Objectives of the organization within the strategic medium- and long-term periods
- Operational goals are the tactical short-term fiscal year performance indicators required of departments and units
- Departmental/unit goals are established within the authority of the corresponding managers to ensure that both the strategic and tactical organizational goals can be achieved with no hindrance

1.2.6.1 Hierarchical Levels of Strategy

Frequent reviews and reassessments of goals are to be conducted according to the volatility or stability of the organization's operating environment. Reviews must be aligned to the timeframe utilized for the goal to act upon the evolving changes that occur in that period.

Strategy formulation is performed to establish goals at all the levels of the organization, from executive management to the operational level.

Strategy can be formulated on three different levels:

- Corporate-level strategy – concerned with the selection and coordination of businesses in which the organization should compete
- Business unit–level strategy – concerned with developing and sustaining a competitive advantage for the organization's products and services
- Functional-level strategy – related to business processes and the value chain in marketing, finance, operations, human resources, and R&D

1.2.7 The Organization's Strategic Path

Change Management initiatives and programs and projects are bound to the organization's stated Vision and Mission.

- Vision statement – *What* I would like to *be*: a description of an organization's aspirations to be achieved or accomplished in the mid-term or long-term future
- Mission statement – *What* I would like to *achieve*: a written declaration of an organization's purpose and focus (Figure 1.7)

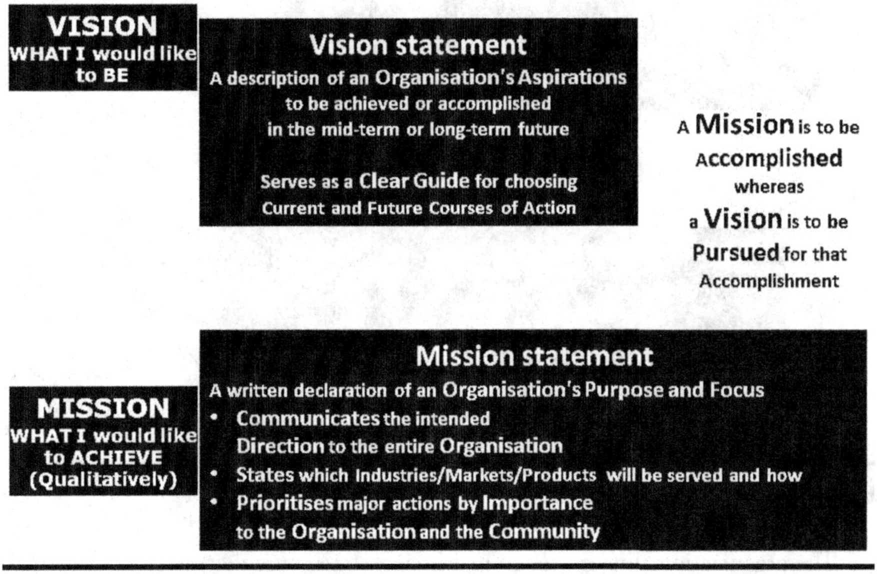

VISION WHAT I would like to BE	Vision statement A description of an Organisation's Aspirations to be achieved or accomplished in the mid-term or long-term future Serves as a Clear Guide for choosing Current and Future Courses of Action

A **Mission** is to be **Accomplished** whereas a **Vision** is to be **Pursued** for that **Accomplishment**

MISSION WHAT I would like to ACHIEVE (Qualitatively)	Mission statement A written declaration of an Organisation's Purpose and Focus • Communicates the intended Direction to the entire Organisation • States which Industries/Markets/Products will be served and how • Prioritises major actions by Importance to the Organisation and the Community

Figure 1.7 Vision and Mission.

To achieve organizational goals, management should establish specific and measurable Business Benefits results to attain. The goals should be time-bound and realistic and should involve:

■ Instituting a performance management system to monitor and control work performed in projects and action plans
■ Assigning appropriate budgets to projects and action plans
■ Setting organizational goals following the "SMART" model – specific, measurable, achievable, realistic, and time-bound

Difference between *Goals* and *Objectives*:

■ Goals and Objectives are derived from the organization's Vision and Mission
■ Goals and Objectives both involve forward motion

Goals are generically for an achievement or accomplishment for which certain efforts are made.

Objectives are specific targets within the general goal.

Goals are the big picture; Objectives refer to specific target plans, each being relatively short-term in nature.

Objectives will be crystalized into nascent initiatives and subsequently broken down at the strategic level of change avenues to pursue, which are then detailed into tactics in a top-down approach.

Figure 1.8   Launch of initiatives.

The organization thus moves from Mission statements that describe a qualitative view of things to quantitative objectives, and these, broken down at a strategy step, eventually lead to tactics, which will be where organizational initiatives are transcribed into programs and projects (Figure 1.8).

1.2.8 From Strategic Vision to Projects

As described above, the framework for the Management of Change by programs and projects is composed of multiple levels. The top-down approach develops the overall scheme of things, translating the different steps to formulate the programs and projects that are to be launched to comply with the organizational goals. However, a feedback mechanism must exist to modulate the strategic direction(s) chosen, as projects are analyzed and detailed, and thus will allow for strategies to be modified where necessary (Figure 1.9).

1.2.9 Organizational Focus

Organizations need to stay focused on the declared Vision and Mission. This must be translated into a coherently defined and executed strategy and action plan. Effective strategic planning requires a dynamic, methodical process that maintains the organization's focus on the right issues and actions. Organizational management must define and continuously redefine four essential components of the strategy and tactical actions:

- Strategic goals – brief statements of what needs to be achieved in terms of operational sustainability, growth, products and services development, market presence, profitability and, most importantly, the retention of skilled employees

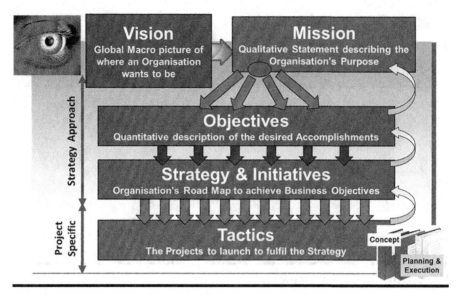

Figure 1.9   From Vision to projects.

- Improvement objectives – specific and measurable performance improvements set within specified timeframes and aligned to specific strategic goals
- Action plans – translating Objectives to corresponding programs and projects, with defined responsibilities, schedules, and cross-functional teams to achieve the Objectives
- Performance measures – providing quantitative measurable results on strategic goals, improvement objectives, and process performance and allowing for the review, evaluation, and update of the originally planned actions.

These four components allow an organization to manage strategically, and to formulate Objectives and action plans that align with strategic goals. Measurement of process performance results will enable the organization to evaluate the need for changes and evolution.

1.2.10 Programs and Projects as Instruments of Organizational Change

The organization will be performing different programs and projects across its structure (Figure 1.10).

1.2.11 Balancing CAPEX and OPEX

CAPEX are the Capital Expenditures used to acquire assets or improve the useful life of existing assets and are made over a period greater than a year. The CAPEX budget

Fulfilling Business Goals by :

➢ TRANSFORMATIONAL PROGRAMMES: STRATEGIC – Organisation wide
➢ OPERATIONAL PROJECTS: TACTICAL – Business Unit & Local

Business Drivers
➢ Transformational
➢ Operational

Change

Business Case

**Programmes & Projects
are Launched to Fulfil Goals**

Providing Business Value and meeting KPI Criteria

Figure 1.10   Fulfilling business goals.

accounts for the potential profitability of the plans involved, adjusted to net Present Value or Internal Rate of Return. The budget will determine the capital needs of the organization and identify the maintenance of the CAPEX expenditures.

CAPEX programs and projects are transformational.

An OPEX budget is an Operational Expense for ongoing costs for a business, product, or system made within a period of one year. The budget is related to the production of goods and services and includes wages for employees, research and development, and costs of raw materials.

OPEX programs and projects are tactical.

1.2.12 *Planning Business and Organizational Change – Leadership*

The organization's Strategic Transformational and Tactical Operational Change Management will demand organizational leadership to conduct the required planning and structuring of the company and the performance of various monitoring activities (Figure 1.11).

The leadership will follow a people-centric approach that guides the alignment of individual goals with strategic goals and Objectives. Organizational leadership is a role that extends beyond the regular management function to motivating employees at various levels to contribute to the achievement of organizational goals, through the implementation of innovative strategies and retaining a productive workforce.

Striking a balance between an employee's development needs and the organization's strategic goals is one of the crucial elements of leadership. The right implementation of such skills enables an organization to retain its productive talents and

Figure 1.11   Change management organizational leadership.

motivate them to contribute to organizational growth. Organizations that practice such leadership skills treat their employees as vital assets.

Organizational leaders inspire and motivate other employees to work together in the right direction and contribute to organizational success.

Without a focus on organizational leadership, Strategic Transformational and Tactical Operational Change Management will be difficult to achieve.

The strategy implementers' challenge in performing these tasks is to bring the organization's conduct of internal operations into alignment with the Organizational Strategy and to unite the organization for strategy accomplishment.

Strategy-supportive structures are needed with organizational skills and capabilities, functional area activities, organization roles and responsibilities, reward systems, and incentives, policies, and procedures, technology infrastructure and information systems, performance measurement mechanisms, budgets, and programs, and most of all, shared values, and cultural norms.

1.3 Transformational Strategies

The transformational strategic projects, which will be CAPEX funded, will be revolutionary in that they will produce new ways of operation for the organization.

Long-term organizational goals are established to convert a Mission statement from a broad Vision into more specific plans and projects. They set the major benchmarks for success and are designed to be measurable, specific, and realistic translations of the Mission statement that can be used by management to guide decision-making.

The degree of change to be expected in the organization's performance will be medium to large, ranging from the introduction of a new product/service to the upheaval of the organization's structure itself. The Return on Investment for these projects, to be realized in a medium-term or long-term period, is a major driver for the programs and projects.

1.3.1 Approach to Implementing Transformational Strategy

There are six principal tasks that constitute the approach to implementing a transformational strategy.

- Engaging in strategic leadership – strategic leadership enables securing employee commitment to the strategy and its accomplishment
- Ensuring the organization's culture fits the strategy, where performance contributes to the accomplishment of the strategy
- Building an organization capable of executing the strategy – the organization must have the necessary structure to convert the strategy into reality. Furthermore, the organization must possess the various skills required to execute the strategy successfully. Related to this, roles and responsibilities must be clearly defined for key implementation results
- Establishing appropriate funding to support the transformational strategy – to accomplish the strategic objectives, management must provide budgets that are adequately construed to fund the human resources, equipment, materials and facilities, and other resources to perform the strategic plan. Further, formal plans must also be developed, with identified financial and temporal targets to be achieved
- Implementing internal support systems, such as policies and procedures, to ensure organizational workforce performance behavior – other internal systems must provide critical information on a timely basis, from operational performance, materials management, customer service, financial accounting, and other systems essential to the organization to enable its strategy-executing capability
- Determining and deploying rewards and incentives linked to Objectives and strategy – business units, departments, and employees must be inspired and motivated to accomplish the transformational strategy

1.3.2 Transformational/Strategic Objectives and Goal Setting

Initiatives launched must frame the Objectives that are to be set for the programs and projects. Tangible results and Deliverables are to be defined, as well as the scope boundary of the program or project. This boundary must incorporate all the functions the program or project must fulfill. The key areas to consider within the boundary are the changes to be instituted in the internal workings of the departmental units; existing processes; the re-skilling of employees and dispensing the required training to accomplish work in the new operating environment; and the technology infrastructure.

1.3.3 Transformational/Strategic Changes to the Organization

The transformational strategic objective will set drastic and significant changes within the organization for its short-term and long-term viability. To achieve the organizational goals, measurable change objectives are to be achieved and will apply across the whole organization, covering business units, departments, functions, and geographies. People, processes, and technology will be affected by the change objectives.

The organizational goals must be described as measurable change objectives to be achieved and applied across the whole organization, including business units, geographies, and functions.

Consequently, the change objectives are scoped in the program or project and will refer to:

- Goal – measuring the goal achievement
- Purpose – change expected in the performing organization with respect to people, processes, and technology
- Outcomes – specific evaluation of operational performance post implementation
- Structure – roles and responsibilities, aligned to a resource budget and schedule

The above elements are to be further described and detailed to ensure conciseness in the project's intended goals:

- Organizational objective the project supports
- Change expected by producing outcomes
- Measures of goal achievement in the budget, schedule, quality, quantity, and other pertinent measures according to the objective set
- Success conditions expected at the end of the project
- Description of specific completed results to be produced by the performing project and team, with a clear description of the completed Deliverables targeted
- Major activities, resources needed, and responsibilities
- Funding and schedule

The launched program or project must therefore ensure that its scope is robust and includes all the necessary functions to achieve change success. Organizational impacts must be identified, and project functions addressing these must be added within the project planning exercise. The range of impacts will vary from project to project; however, the major areas to consider are:

- Organization structure
- Culture
- Policies and procedures
- Communications
- Human resources
- Job responsibilities
- Motivation
- Reward systems

This is further developed in Section 1.5.

1.3.4 Organizational Change Competency

Strategic Transformational Change cannot be successful unless the organization addresses the present competencies and those required for the future within its workforce.

For any initiative destined to be launched, the program/project scope contents must include actions to ensure that employees are satisfactorily trained and re-skilled. The business analysis must develop a competency change scheme following a current-state assessment and resulting gap analysis, and the change and improvement program/project will subsequently define, design, and plan for the implementation of the education and training curriculum programs.

Competency changes to be implemented will be defined in specific Deliverables, organizational structure changes to be realized, hiring or outsourcing requirements, and funding and transition schedules to be adhered to, with specific attention to the current state of:

- Organization structure
- Culture
- Policies and procedures
- Internal and external communications
- Human resources' short and long-term plans
- Roles and responsibilities
- Performance recognition schemes and reward systems

1.3.5 Institution of a Measurement Mechanism

Performance measurement processes exist in all organizations. However, challenges exist in the relationship between strategies driving transformational and Tactical Operational changes and the collected organization-wide measured performances.

There may be an accountability disconnect at the operational level, as the operating units have a greater focus on daily performance analysis, rather than strategic and transformational management decision-making. Effective measurement efforts are further hindered when the organization employs disjointed and outmoded systems, while collecting an overwhelming number of metrics.

To accomplish strategy execution and performance improvement, an organization-wide performance measurement process is to be instituted and utilized. This process will provide a mechanism to monitor progress against transformational and tactical goals. All organizational units can thus interact transparently and effectively, crossing structural silos.

The performance measurement process will ensure the alignment of strategic objectives with day-to-day operational goals. A reporting mechanism of metrics, timely and frequently collected at the program/project levels and from the day-to-day operational performance, will allow for strategic and operational decisions to be effectively made to respond to progress changes or modifications against originally set goals.

The performance measurement process would provide for measuring important business processes, information about the effectiveness of these business processes, and the ultimate results of these processes. The process is to be further accompanied by the collection of internal measures such as revenue, operating income, cash flow, and asset utilization. External measures would also be required such as cost comparison vs. competitors, capital redeployment techniques used by competitors, innovative approaches to product distribution, and measurements of customer complaints.

1.4 Tactical/Operational Strategies

Tactical strategies are those that address day-to-day organizational unit activities to sustain, maintain, and improve operational performance. Tactical strategies will also be subordinated to executive decisions that concern the organization's market share, competitive pricing, customer service, or other facets of overall performance. Tactical strategies are limited to the performing operational units and are short term in nature as they would be achieved within a fiscal year.

1.4.1 Sustaining Operations

Tactical Operational projects, which will be OPEX funded, focus on the ongoing organization's performance and will be more evolutionary. The degree of change will range from medium to low, as the operating units will be launching projects with a variety of scopes from "fix as fail" to "continuous improvement". The Return on Investment for these projects, usually to be realized within the fiscal year, is driven by operational goals aimed at sustaining and maintaining operational performance levels.

Whether programs and projects are transformational or tactical, the organization will need to evaluate thoroughly the change impacts on its operational processes and the constituent procedures of its products and services. A lack of change impact assessment will affect the organization's ability to meet its goals unless a comprehensive Organizational Readiness project is also performed.

1.4.2 Operational and Departmental Goals

Operational goals are short-term tactics designed to achieve a company's long-term strategy. An operational plan is subsequently established which provides the envisaged project with the contents of the required scope. The plan will describe the specific goals to be met; the major actions to perform to achieve the required results; the human and physical resource availability; the financial and temporal constraints; and any other internal or external constraint specific to the project.

As operational goals are intended to address the sustainability and maintainability of the day-to-day activities of the organization's performing unit, goals are to be sufficiently detailed and measurable, with an important consideration of the change's impact on current operations.

1.4.3 Planning Operational Objectives Effectively

The very essence of planning operational objectives effectively is a continual process of ensuring that operations are constantly and consistently doing the "right things right".

Operations strategy and overall business strategy are inextricably linked. While a well-defined operations strategy will not guarantee success by itself, not having one will almost certainly guarantee failure. An operations strategy and the requisite skills to develop and implement that strategy are critical for achieving the operational performance objectives.

Operational entities concentrate on their ability to perform according to measurable operational criteria that have been set for the actual calendar and fiscal year.

This process involves continuously reassessing current performance strategies, including Objectives, action plans, and measurements. It also involves developing new or modified objectives, action plans, and measurements whenever needed. Effective strategic operations planning must continually and systematically perform the following two tasks:

■ Reassess the current strategies, Objectives, action plans, and performance measurements, and examine how well they are reflected in the company's overall strategic plan
■ Develop new or modified objectives, action plans, and performance measurements that are well-connected to the overall strategic plan

1.4.4 Operational Planning for Projects

Operational planning for projects primarily addresses tactical strategies of a given organizational unit and would consist of those projects destined to maintain, sustain, and improve the process of the operational performing unit.

Operational planning, however, will also be required to link the organizational strategic goals and Objectives to the unit's own tactical goals and Objectives. The high-level transformational initiative describes how a strategic plan (or a defined portion of a strategic plan) will be put into that unit's operation during a fiscal year or another given budgetary term. Thus, operational planning will not only determine the projects required for its specific organizational unit but will also integrate projects within the unit's domain that are driven by and or imposed by a Strategic Transformational Initiative.

The interaction of operational tactical projects and those derived from transformational initiatives will often lead to counterproductive situations, whereas previously set operational goals may not be achieved due to higher-order organizational goals. Additionally, the operational unit will be confronted with resource conflicts, as those will be required to be shared between the unit's needs to maintain and sustain its operation, its specific tactical projects to be performed in the fiscal year, and the transformational initiatives.

It is therefore imperative that both tactical and transformational goals be synchronized and an assessment be made of the organization's ability to successfully perform the intended projects.

1.4.5 Planning for Changes in Operations

A prerequisite for action plans is to develop specific improvement objectives tied directly to the organization's Business Goals.

The operational performance of the organization is the most important factor in its viability and success. Irrespective of whether a tactical project is performed or is issued from a transformational initiative, the implemented delivered results must prove to be beneficial to the organization's overall operational performance. Therefore, all projects performed in the unit's organizational domain must be assessed for the changes that will occur in its operation.

These projects will cover a variety of operational objectives of sustainability and process improvement and will all focus on different aspects of performance:

- Cost: to produce at a low cost
- Quality: to produce in accordance with the specifications and without error
- Speed: to respond rapidly to customer demands
- Dependability: to deliver products/services in accordance with contracts
- Flexibility: to change operations

It is essential that the operational strategy includes Organizational Readiness actions in all the projects it launches. These actions cover re-skilling, retooling, and the transition from old to new processes.

1.4.6 Developing Operational Action Plans

A specific action plan defines the sequence of time-phased activities or tasks and the resources necessary to reach the desired project objective. In addition, each activity and task required to achieve the objective has an accountable individual assigned. The action plan responds to the project scope contents and is bound within a timeframe with clear lines of reporting.

The action plans must integrate a comprehensive situational analysis of the operational environment and include:

- Analysis of business processes and events
- Determination of resources requirements
- Establishment of dependencies between processes and events
- Analysis of information on the effectiveness of the plan for the eventual operational performance

1.4.6.1 Conducting the To-Be Process Design Phase

The purpose of this phase is to define and document the optimized To-Be processes, eliminating the identified current business process weaknesses that are ineffective or need improvement. This will include, but is not limited to:

- Developing initial hypothesis on improved process
- Involving key business process owners in the redesign team
- Testing the hypothesis with a cross-functional group of Stakeholders and refining
- Communicating results and generating stakeholder commitment

Diagrams of the "To-Be" processes such as use-case and activity diagrams and data models are to be developed, and a draft map of the improved process is to be created in a final report with recommendations.

1.4.6.2 Competences/Skills Reviews

The operational unit must draft an organizational skill development plan and a plan for the use of organizational resources to improve efficiency and expand productivity in readiness for the new environment. The results of these reviews must be integrated into the project scope contents.

1.4.7 Executing Operational Change Plans

The major challenge to the operational unit in the execution of its change plan is addressing the concomitant development and implementation of the change project while maintaining current operational performance.

The operational change plan execution must consider which resources are required to be assigned to tasks to be performed for its completion and how these utilized resources must be allocated so as not to deplete the operational workforce and without interfering with the unit's current operational performance capability. Operational change projects will be successful only if the operating environment can transition to its new state effectively, depending on the preparations made for Organizational Readiness, and included in the project scope contents.

1.4.7.1 Organizational Readiness

Readiness means being prepared. Organizational change readiness ensures the establishment of:

- The right conditions and resources in place to support the change process
- A clear Vision and Objectives for the intended change
- The motivation and attitudes to engage with the change and make it successful

Of utmost importance are the organizational transition period of the change project and the actions undertaken to address the unit's Organizational Readiness by the analysis of the level of preparedness of the conditions, attitudes, and resources, at all levels. The greater the complexity of the proposed change, the greater the importance of understanding where Organizational Readiness is critical. The scope of the proposed change will impact a range of process and system components and challenge any or all the elements within it for their transition readiness.

When determining readiness for change, the culture and history of change in the organization must be considered, for accountability, resource availability, and availability of staff with Change Management knowledge and experience.

Organizational Readiness must consider the human side of change, by supporting the people and culture involved in the change. A comprehensive understanding of the current organizational processes will determine the migration path to follow from the "current state" to the "future state" processes, from process improvements, from increased efficiency in daily operations, and from reskilling staff and retooling where required.

1.5 From Strategies to Initiatives and Projects

Please also refer to Chapter 2, "Management of Programs", where this topic is expanded upon.

All transformational and tactical strategies will generate initiatives and programs/projects.

As programs and projects are initiated and launched, and for overall goals to be achieved, organizations must ensure that the operational departments and units have well-defined and agreed-upon Business Goals, operational objectives, action plans, and performance measures.

Management must diligently define and redefine the essential components of a successful strategy and tactical actions by deploying programs and projects that maintain the organization's focus on the right issues and actions.

While corporate strategy seeks for the organization to sustain and flourish within its environment over the long term, operational strategies revolve around the decisions and actions taken in the short term which have a direct impact on the organization's ability to fulfill its strategy.

To meet its tactical performance objectives, operations will launch programs and projects of continuous improvement and TQM with:

- Improvement objectives: specific and measurable performance improvements set within certain timeframes and tied to specific goals
- Action plans: Objectives translated into a specific set of steps, responsibilities, schedules, and cross-functional teams for implementing the plans to achieve the Objectives
- Performance measures: quantitative means of reviewing, evaluating, and updating actions, improvement objectives, goals, and process performance

Operational programs and projects will focus on cost, quality, speed, dependability, and flexibility:

- To produce at a low cost
- To produce in accordance with specifications and without error
- To respond rapidly to customer demands
- To deliver products/services in accordance with demands
- To adapt operations to different rhythms

All transformational and tactical strategies will generate initiatives and programs/projects.

1.5.1 Nature of Initiatives, Programs, and Projects

Programs and projects, irrespective of whether they are transformational or tactical, will for the most part focus on any or all of the following strategic directions: competence-based; capability-based, and resource-based strategic directions.

The scope of the program or project will thus deal with a wider and broader perspective.

1.5.1.1 Competence-Based Strategy

Competence-based strategic management incorporates economic, organizational, and behavioral elements in a framework that is dynamic, systemic, and holistic. The strategy will expand the program or project scope to include actions to:

- Invest in human capital; build skills in people at all levels
- Broaden the technical, problem-solving, decision-making, and leadership skills of those who are in the front line
- Make skill-building a key performance measure for all employees
- Strengthen the facilitation, managerial, listening, delegation, communication, and diversity skills.

The strategy for the program or project will be to incorporate the ability to sustain the coordinated deployment of resources in ways that help an organization achieve its goals of creating and distributing value to customers and Stakeholders. The program and project will also develop the education and training plan as a major component to facilitate the organizational transition.

1.5.1.2 Capability-Based Strategy

A capability-based strategy is a specific ordering of processes, people, resources, and information and technology aimed at creating a defined business outcome.

The launched program or project will be the bridge between the "As-Is" and "To-Be" states.

1.5.1.3 Resource-Based Strategy

A resource-based strategy emphasizes building around the further exploitation of existing core competencies and strategic capabilities.

The launched program or project will strive to incorporate the effective and efficient application of all useful resources that the organization can assemble to create and maintain its competitive advantage. These cover knowledge, skills, motivation and collaboration, and tangible financial and physical resources.

1.5.2 Implementation of Change

Strategic objectives translate the organization's directions. Operational objectives are developed by business units and functional units and focus on the operational performance of the organization. A major challenge exists when both sets of Objectives are competing within the same timeframe, same resources, and for similar sources of funding. Programs and projects must be synchronized within a given timeframe as they will be overlapping in areas such as product development; supply chain management; marketing and sales; human resources; financial management; and information technology.

The organization must avoid being faced with change functions that are diametrically opposed to one another and operate in a way that makes meaningful business performance improvement nearly to completely impossible.

1.5.3 Organizational Impacts

The scope of the initiative and subsequent program/project and the associated impacts will affect the organization's ability to meet its goals (Figure 1.12).

Conducting an Organizational Impact Analysis determines the interrelated conditions that may make the proposed initiative unfeasible.

The impact analysis will cover the following major domains: all internal and external operating procedures; organization culture and organizational roles; staffing with the corresponding necessary skills and training where required; and reporting structure.

Impacts to process and product/service changes have to be assessed, covering all derivative projects in operations for additions/augmentations to current performance, platform projects for next-generation products/services, breakthrough projects with ambitious Objectives, and development projects for new products/services.

All domains are to be considered within the scope of the initiative and subsequent program/project (Figure 1.13).

1.5.3.1 Process Improvement

Process improvements of all types will be performed in both the transformational and the tactical areas of the organization. The programs/projects launched will

Figure 1.12 **Organizational impacts.**

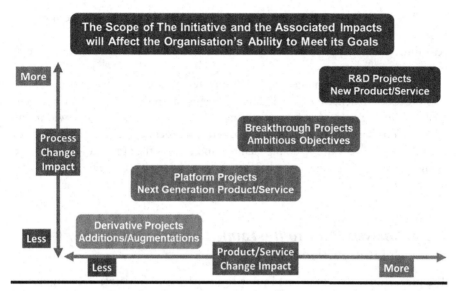

Figure 1.13 Impacts on processes and products/services.

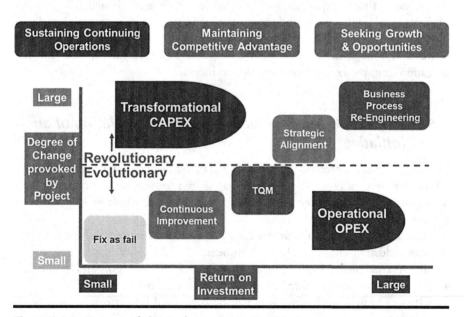

Figure 1.14 Degree of change by projects.

generate a degree of change that will be either evolutionary, for tactical change, or revolutionary, for transformational change.

Tactical changes, such as "fix as fail"; continuous improvements; and TQM-driven changes will, for the majority of cases, be programs/projects of a tactical nature and that are OPEX funded (Figure 1.14).

While transformational programs/projects of redesign, reengineering new product/service creation will be CAPEX funded, these are not restricted to a fiscal year timeframe. A special impact analysis is to be conducted for business process redesign.

Business Process Reengineering (BPR) involves the radical redesign of core business processes to achieve dramatic improvements in productivity, cycle times, and quality. Thus, programs/projects of this nature must cover a wide scope within the organization that covers not only the most critical processes but also the impacts of re-skilling employees and retooling facilities and ensures the organization's readiness by the implementation of the work to be performed to facilitate the transition to the new operating environment.

1.5.4 Analysis Prior to the Launch of an Initiative and Its Related Project(s)

The preparatory work performed prior to launching an initiative is accomplished by conducting a business analysis and preparing a Business Case. This is a project in itself. This is expanded and detailed in Chapter 3, "Benefits Realization Management". Suffice it to understand in this chapter that these enabling documents must be established to validate the essence of the initiative planned to be launched. Of major importance for programs and projects is the focus on how the transition from present to future will be conducted.

1.5.5 From High-Level Objectives to the Formulation of an Initiative

Business analysis is a project, and typically will be conducted at an exploratory stage, followed by an iterative stage of options analysis and refinement, including the evaluation of the proposed direction envisaged, and would conclude with a high-level plan for the intended initiative.

The external analysis will cover the market: current customers and clients; competitors; industry trends; and the environment.

The internal analysis will focus on the organization's strengths and weaknesses and its current employee competencies and perform financial modeling.

Options analysis and refinement will be conducted by performing iterations, including detailed financial modeling, gap analysis, and feasibility, with a view to establishing the most appropriate initiative to pursue.

The high-level plan for the selected initiative will establish the overall scope, major actions, timescales, funding, resources, and constraints that will drive the parameters of the to-be-launched program or project (Figure 1.15).

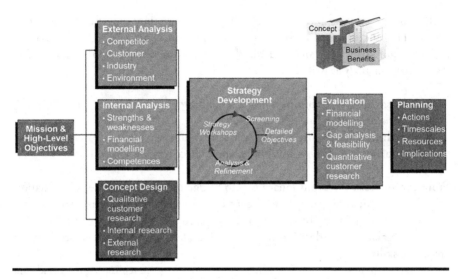

Figure 1.15   Business analysis framework.

1.5.6 Performing "As-Is" and "To-Be" Analysis

Business analysis will begin with an analysis and an understanding of the Current Situation, the "As-Is", with the assessment of the internal operating environment as well as the external forces that exist and the valuation of the impacts that the initiative and subsequent project will have on the current organization (Figure 1.16).

According to the breadth of the scope of change across the organization, the As-Is will assess the current state in the different domains and departments affected,

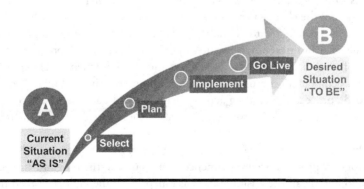

Figure 1.16   "As-Is" and "To-Be" change context.

such as manufacturing/production/distribution, marketing/sales, finance/administration, HR, and information systems. All business processes within the scope of the change required are to be analyzed and their current methods of work assessed. The As-Is situation analysis will:

- Perform analysis of business processes and events
- Determine critical business processes
- Establish dependencies between processes and events
- Complete data analysis and information effectiveness

From the analysis, the evolutionary gap to bridge to the "To-Be" desired situation can be determined. Similarly, the current organizational structure is analyzed and reviewed to ascertain the breadth and depth of change to be exercised for reporting and escalation for decision-making and underscoring the transition gap in employee skills.

Situation analysis is further conducted by reviewing the technological backbone of the enterprise as well as its physical infrastructure and facilities (Figure 1.17).

The To-Be-desired situation reflects the organizational goals and the results sought, as defined in the Objectives that have been established. However, the To-Be goals remain at this stage aspirational and must be converted to tangible results to validate the need to launch a program or a project.

The To-Be design phase defines and documents the optimized To-Be processes, eliminating identified current business process weaknesses that are ineffective or need improvement.

Of major importance will be the effect on the organization's operation and the need to enhance employees' skills and competence. Based on the s-Is analysis, employee training and education plans are to be identified to be implemented. The To-Be design must establish how the organization will transition the capability of the performing staff from its current state and how the operational entity must prepare its Organizational Readiness for the new environment to achieve improved efficiency under the changed environment.

The phase is conducted to develop initial hypotheses on improved processes and must involve key business process owners in the redesign team. Hypotheses are tested with cross-functional groups of Stakeholders and are subsequently refined. The To-Be design documentation must be robust and comprehensive and include diagrams of the To-Be processes, including activity diagrams and data models, accompanied by a draft map of improved processes. A final To-Be report with recommendations is to be completed.

The business analysis would establish a high-level road map that will be detailed by the program/project once it is launched. Subsequently, the program/project plan will expand and detail the business analysis blueprint of the high-level road map by

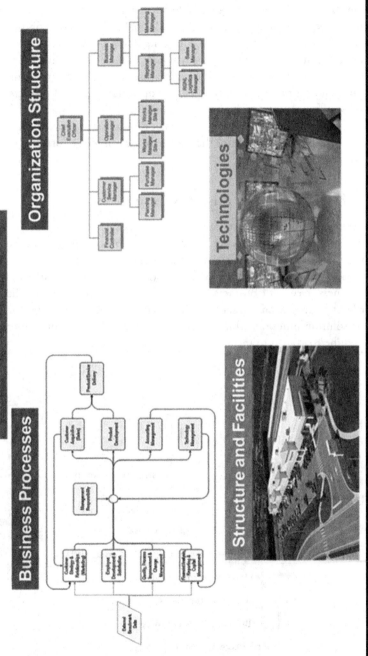

Figure 1.17 Processes and structure impact analysis.

performing the project's front-end engineering and design. Decisions can then be made to determine if the To-Be goals can be realistically achieved.

A major consideration to be apprehended will be the anticipated timeframe imposed or requested to accomplish the To-Be goals. The duration of the timeframe will affect how the project will fare during its development cycle. The longer the timeframe the more "potential change will exist in the change", and that will be the case for the majority of Strategic Transformational projects. Shorter timeframes of less than a month or a quarter, typically for operational tactical projects, however, will be less affected and may have fewer change impacts.

The s-Is and To-Be analysis stage is of major importance as it will set the boundary of the initiative and influence the choice of the number of projects necessary to accomplish the goals set.

1.5.7 Establishing the Business Case

The development of the business case is a mandatory step prior to any project launch. The more the business need is pertinent to sustainability and growth, the more the Business Case is to be comprehensive. Business Cases can range from a one-pager to a full and all-inclusive documented analysis of current and future situations that meet business needs. All Business Cases, however, must include the following two major facets: the Current Situation, which states the need, issue, and/or problem, and the solution proposal, which presents the desired business outcomes and benefits, and the road map to meet these (Figure 1.18).

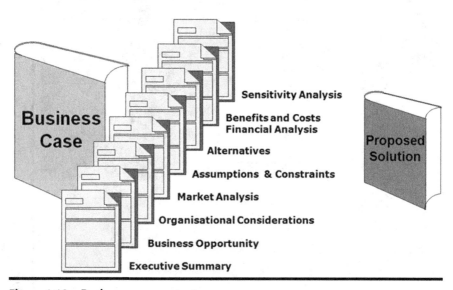

Figure 1.18   Business case contents.

There is no hard and fast rule for the structure of the Business Case; however, all Business Cases must include the following:

1. The *executive summary* highlights the key points in the Business Case. It states the business problem by providing a thorough and objective presentation of a business issue that requires a timely solution. The summary must highlight the benefits and the Return on Investment. It must also demonstrate how the Business Case responds to the corporate strategic and/or operational alignment
2. The *business opportunity* describes the incentive for the project that the Business Case will describe and propose. The business opportunity includes a definition, a statement of scope, and a discussion of Objectives that the project will help the organization achieve. It will explain how the expected outcome of the project supports and aligns with the organization's strategic or operational intent
3. The *organizational considerations* examine the current organization and highlight the areas where the proposed project will affect the future structure
4. The *market analysis* examines changes in the business environment, such as clients, competitors, suppliers, industry standards, and legislation
5. The *constraints* are internal and external. They will cover the current and future constraints originating from schedule, resource, budget, staffing, technical, and other limitations that may affect the success of a project
6. The *alternatives* section analyzes the different choices that can be considered and their respective merits
7. The *assumptions* are events on which the proposed solution of the Business Case is based. Assumptions must be validated for a project to succeed
8. The *benefits and costs* provide a comprehensive assessment of both expected benefits and anticipated costs throughout the total Product Life Cycle. These benefits are to be qualified in terms of their impact on the organization's operational performance
9. The *financial analysis* compares benefits to costs and analyzes the value of a project as an investment. The analysis must be credible on the Return on Investment and Payback Period, by identifying the ownership costs required for the project development and the operational costs for the utilization of the end product or service. Projects of an anticipated duration greater than one year must also consider the net Present Value and Internal Rate of Return. The financial analysis may also need to include a cash flow statement
10. The *sensitivity analysis* evaluates the project risks and presents the mitigation and contingency plans and the proposed solution's risk exposure
11. The *proposed solution* should offer an approach to solving the business issue/problem and demonstrate that the project is both viable and beneficial. The proposed solution must emphasize the merits it has over the alternative solutions that exist. The proposed solution should address only the agenda described by the business problem, and the approach narrative should tie the solution to the existing business problem in cost/benefit terms

Project Objectives should be explicitly stated and be aligned with the organization's strategic and operational plans and infrastructure. The project's key Stakeholders must be referenced as well as the nature of sponsorship and upper management support and commitment.

An Organizational Impact Analysis should describe the anticipated impact of the project on the organization's people, processes, and technology.

Budget/resources estimates (mostly in order of magnitude) should consider all cost components of the ownership and operations cycle, including support, maintenance, and other recurring costs.

A high-level project implementation plan is to be established spanning the initiation to close-out phases. Any plan at this stage can only be tentative and must be considered preliminary. The subsequent initiation and planning phases will refine the plan. The proposed project development plan will include:

- Deliverables schedule
- Phase/stage definitions
- Macro activities
- Workload estimate/breakdown
- Project schedule
- Required resources
- Funding requirements
- Project leadership team, project governance team
- Project controls and reporting processes

The recommendations summarize the main points of a Business Case and offer suggestions on how to proceed with the project.

1.5.7.1 Validation and Approval of the Business Case

The business case is reviewed and validated, and the choice of the proposed solution is assessed. Foremost, the proposed project is gauged for its consistency with the organization's strategic and operational plans. It is also appraised for the value it will create, and whether it is within the investment and budget guidelines.

A formal project origination process is important because it creates a focal point for the assessment and analysis of all proposed initiatives, directs resources to the Right Projects, and focuses the organization on projects that will deliver the greatest value. Each organization has its own project approval process. However, this process revolves around three major steps:

i. Develop project proposal – the business case is established, and the initial project boundary and parameters are defined
ii. Evaluate project proposals – quantitative financial analysis is performed, such as cost/benefit analysis and ROI, and the proposed projects are evaluated against a set of pre-established strategic and operational business criteria

iii. Select projects – following agreement on the project's feasibility, this is ranked with other proposed projects and a decision is formally made regarding the project proposal

To make a meaningful evaluation, the decision-making body must possess sufficient information on the project's business case and the viability of its proposed solution. The competing projects must be evaluated and compared using a consistently applied methodology and the selection process must consider the project's fit with the organizational strategic/operational plans.

Once a project is selected, funding and/or further management commitment is required to progress to project initiation. Formal project sponsorship is then formally established.

The project proposal process in the origination phase may be part of the total budget cycle, serving as the justification for funding requests. In this case, the project proposal is to include a budget estimate for the total cycle and a funding request to proceed to and perform the project's initiation phase.

1.5.8 Organizational Readiness and Resistance

Key to the effective transition from As-Is to To-Be is the Organizational Readiness of the operational departments and units. The receiving entities must implement the required employee training, process adaptations, and infrastructure enhancements for the change to be successful. This will involve establishing and securing:

- Intellectual commitment: the comprehension and belief that change is necessary and the proposed changes are appropriate
- Emotional commitment: the willingness and enthusiasm to accomplish the planned change and "weather the storm"
- Required skills: the competence and capability required to realize the change and operate effectively in the new environment
- Management support: the support necessary to carry out the change and operate effectively in the new environment

Resistance to change centers around how well or badly the existing organization responds to the change and the failure to predict the time and problems inherent in an implementation initiative. Lack of executive and management support and poorly elaborated and forecasted strategies will create confusion leading to the ineffective implementation of change. The lack of focus on people, pressed by optimistic or inflexible implementation of schedules, will generate passive or active resistance and stifle any possibility of creativity and innovation.

Please refer to Chapter 3 – Benefits Realization Management and Chapter 4 – Stakeholder Management and Engagement where these two topics are explained in detail.

1.6 Resistance to Change

The scope of transformational and tactical changes will modify an organization, ranging from a moderate shift to an upheaval in its operational environment.

Initiatives and projects will create change in the way the organization will operate, in the processes used, in the new skills to acquire, and in the size of the organization.

An ill-prepared change transition will create resistance.

Organizational resistance will vary depending on the scope and extent of the change. It will originate principally from threats to competencies and skills, established power relationships, and organizational resource distribution, and may well lead to structural lethargy.

Individual resistance is caused by the way people work, in what people will earn, in the way people will interact, and in the size of the staff. Resistance is principally caused by a shift in the individual's comfort zone, with a feeling of loss and the fear of the unknown. Economic factors judged to be negative and insecurity both intensify resistance.

Resistance to change will be manifested in a variety of behaviors including confusion, immediate criticism, denial, malicious compliance, sabotage, easy agreement, deflection, silence, and potentially "in-your-face criticism". The forms taken by resistance may be overt and immediate, where individuals voice their complaints and engage in industrial action. Resistance that is covert, implicit, and deferred will cause a loss of employee loyalty and motivation, increased errors or mistakes, and more absenteeism.

1.6.1 Development and Deployment of a Change Resistance Plan

A change resistance plan is to be established from the earliest stages of the project covering the domains of people, processes, and infrastructure, where organizational resistance can be addressed by establishing and reiterating the direction of the organization's Vision for the change, and the strategies for achieving the corresponding Goals and Objectives.

For the change resistance plan to be acceptable and credible, groups and people must believe that the change is beneficial and possible and that the timeline to bring about the change in the organization is realistic. Importantly, processes, new skills, and new job definitions must be coherent. In many societies, issues relating to the culture in which the organization operates have to be considered.

Consolidated operational results from measured improvements are to be incorporated to enable reassessment and applying necessary adjustments in the change programs. The effectiveness of changes can be reinforced by demonstrating the relationship between new behaviors and organizational success.

Resistance from individuals is the most important domain on which to focus. Key fundamental principles of behavior have to be covered, such as an understanding that:

- People deal with change to some extent every day
- People going through the stages of change may progress, regress, and then progress again many times
- People may go through several different change processes simultaneously (related to different changes)
- It is possible for an individual not to move out of one stage or to move out of a stage prematurely

In managing resistance to change, it is important to consider the concept of individual adaptability and to incorporate the three main factors that describe how some people are able to effectively adapt to change and stress: commitment, control, and challenge.

- **Commitment**: Where individuals believe in themselves and the work and have a feeling that they are an integral part of the group and feel a connection with others
- **Control**: Where individuals have relative control over events and can influence outcomes
- **Challenge**: Where individuals view change as a challenge and look for opportunities to grow and develop

Planning to overcome individual resistance to change is a multipronged effort that encompasses initially establishing a sense of urgency by creating a convincing reason for why change is needed and removing barriers to encourage risk-taking and creative problem-solving. People need tangible results to measure success, and plans are to be established to create and reward short-term "wins" that move the organization toward the new environment. The recognition of individual changes is strengthened by acknowledging new behaviors and their relationship to organizational success.

1.6.2 Stages to Adjusting to Change

Based on the work of Elisabeth Kübler-Ross, the stages of adjusting to change from the old to the new state are denial, resistance, exploration, and commitment (Figure 1.19).

The change resistance plan is to address these stages for the group and specifically where required per individual in the group. The plan will seek to improve the ability of the organization to adapt to changes in its environment and change the behavior of individuals and groups in the organization.

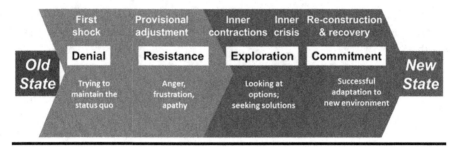

Figure 1.19   Stages in adjusting to change.

The Program/Project Manager is to be the promoter and the driving change agent and act as a catalyst for managing the change created by the project.

1.6.2.1 The Change Resistance Plan – Denial Stage

In this stage, groups and individuals will be trying to maintain the status quo and manifest an initial denial as to why the change is necessary. Those manifesting denial will often use the "*you*" when relating to those who are responsible for implementing the change.

From its outset, the program/project change plan must provide comprehensive information on the scope of the change and the rationale behind it to all groups and individuals concerned. Throughout the development life cycle, the Program/Project Manager is to promote and describe the positive aspects of the change by using a variety of communication methods, such as one-on-one discussions, group meetings, through letters and memos, e-mail and newsletters, posters, notice boards, etc.

1.6.2.2 The Change Resistance Plan – Resistance Stage

In this stage, anger, frustration, or apathy will be manifested by those who have still not fully accepted their denial and/or have understood that the change is necessary but have not yet psychologically incorporated this into their thinking and behavior. Those manifesting resistance will often use the "I" when expressing their concerns.

The program/project change plan is to contain actions to:

- Encourage people to voice their concerns constructively
- Listen to and acknowledge feelings and concerns
- Provide empathy and support
- Communicate without criticizing
- Engage in different ways to move away from the past

By listening to and incorporating remarks and suggestions from the concerned groups and individuals, adaptations to the changes can be considered to be made in the operational environment.

1.6.2.3 The Change Resistance Plan – Exploration Stage

This stage will demonstrate an important shift in the group's and individuals' behaviors, as the "*we*" becomes predominant in the interchanges between the Program/Project Manager and the groups and individuals. The use of the inclusive pronoun signals that the group and individuals are exploring options and seeking solutions.

1.6.2.4 The Change Resistance Plan – Commitment Stage

This stage is when the group or individuals have reached a successful adaptation to the new environment, and operational performance is not hindered by any lagging resistance. It is important to recognize accomplishments, congratulate work performed, and praise commitment. Present and future commitment is reinforced when the organization employs a rational and fair method with tangible rewards.

1.7 Project Portfolio Management and the PMO

Please also refer to Chapter 5, "Project Portfolio Management (PPM) and the Project Management Office (PMO)", where this topic is expanded upon.

1.7.1 Project Appraisal

Projects must originate with a clearly stated business intention. They must also demonstrate how they will create sustainable value for the organization.

A formal project appraisal process is important because it creates a structured approach to assess all proposed initiatives, avoids allocating resources to lower-priority projects, and maintains the organization's focus on projects that will deliver the greatest value.

The purpose of project appraisal is to provide a mechanism for recognizing and identifying potential projects within the organization, to evaluate projects proposed for the next investment cycle, and to reach a consensus on projects to be selected. During this phase, the strength of a project's business case is tested, and the viability of the proposed solution is explored against the company's strategic plan and budget guidelines.

Other factors to take into consideration include legislative restrictions, regulations, and HSE requirements. The key components of the project appraisal process are:

- The project must provide sufficient information about the viability of the project's business case and the feasibility of its proposed solution
- Projects must be assessed, evaluated, ranked, and prioritized using a consistently applied methodology
- The process must consider the project's fit with the organizational Mission and strategic plan

In the project appraisal stage, many projects will contend for limited funding and limited resources. Each project is to be assessed for its alignment with the strategic or operational corporate goals. Evaluation of the proposed project is made against the provided enabling documents. The major documents mentioned below are assessed to ascertain the project's alignment with the business.

- Current Situation Analysis "As Is" – assessment of the ongoing and current operation following the analysis of the business processes with their owners. The boundary of the project is defined and issues to address and resolve are identified
- Desired future state definition "To Be" – description of the desired future state, following the conduct of a gap analysis, identification of a road map to reach the desired future state, and creation of a portfolio of key projects to perform, supported by financial data, including ROI analysis and NPV
- Interaction with the business strategy – definition and implementation of a formal and purposeful communication program, developed with business process owners and line-of-business decision-makers
- Evolved organizational infrastructure model – description of the relationship between organizational infrastructure, business processes, and information management needs

1.7.2 Projects and Project Portfolio Management

Projects contend with each other owing to limited and constrained funding and resources. It is essential for any organization to establish a solid and pragmatic process for selecting those projects that would bring the most value and reap the best benefits.

Each organization will choose how best to approach the management of projects and the level of discipline it wishes to follow to select projects that will fulfill the strategic and operational goals. A haphazard and/or subjective selection technique often leads to missed opportunities and a waste of funding and resources. An objective and informed project selection process does not guarantee total success; however, it will improve the chances of success.

Whatever selection process is followed, there must exist a mechanism that centralizes and brings together all project contenders, to be assessed and compared, and from which the performing organization can determine the priorities to allocate to each project. This mechanism has many names; however, the most applicable and efficient one is "project portfolio".

PMI offers a very comprehensive definition:

A portfolio is a collection of projects and/or programs and other work that is grouped together to facilitate the effective management of that work to meet strategic business objectives. The projects or programs (hereafter referred to as "components") of the portfolio may be mutually independent or directly related. At any given moment, the portfolio represents a "snapshot" of its selected components that both reflect and affect the strategic goals of the organization – that is to say, the portfolio represents the organization's set of active programs, projects, sub-portfolios, and other work at a specific point in time.

Project Portfolio Management (PPM) groups programs/projects so that they can be managed as a portfolio. PPM ensures that programs/projects and expenditures are aligned with corporate strategy and operational objectives.

1.7.3 *Aligning the Project to the Organizational Strategy*

An organization's Strategic Intent can only be accomplished by an efficient and focused approach of Management By Projects. The Strategic Intent is translated into road maps constituting sets of programs/projects all requiring allocation of resources. Programs/projects are periodically reviewed to confirm their continued alignment with the Objectives, and their priority is adjusted depending on their performance.

For the execution of the strategy, programs and projects will be launched for transformational, business focus, or Tactical Operational departmental goals. Project portfolio management answers the question, "Are we doing the right thing?", the "right thing" being the question of whether the project meets the organizational goals and aligns with the strategies to be achieved.

Without the use of a formal PPM, too many projects may be launched that do not answer that criterium. There would be overuse or misuse of resources, both internal and external, and misdirected use of funding. Additionally, due to the interaction of disparate projects, conflicts arise because of competing and contradictory goals. In situations of this nature, management will be faced with budget overruns, schedule slippages, and unaccomplished transformational and tactical goals and the realization of benefits.

For all launched projects to best serve the interests of the organization, a rigorous selection process must be implemented. A selection process is to be instituted as choices must be made and to achieve organizational goals. Those choices, however tough, need to be made so that what enters the funnel realizes the goals and provides Business Benefits.

Project Portfolio Management is a pivotal process for the successful fulfillment of Strategic Intents. Depending on each organization, the visibility given to the portfolio by all levels of executive and operational management will be a key critical success factor. If selection decisions are not made at the portfolio level, by default the project portfolio is the result of individual project choices made one at a time with little regard for the impact that one project has on the next.

Members of the portfolio executive group or steering committee, sponsors, and key Stakeholders of all programs/projects in the portfolio are to receive pertinent and up-to-date progress/status information on the portfolio performance. Operational management needs to be informed as to the Organizational Readiness requirements they have to prepare to fulfill the Business Benefits after the completion/handover of any subset of a program/project.

Upper executive management is to be informed on the investment/funding requirements for the medium and long terms. In many cases, this is done by a "rolling" quarterly financial statement covering 3–5 years or more depending on the organization.

Upper and executive management must inform the Project Portfolio Management of any anticipated changes in the company's direction to ensure the alignment of programs/projects with the modified Strategic Intent(s).

1.7.4 PPM Operational Cycle

Project Portfolio Management (PPM) is a framework that translates strategy into programs/projects and aligns these with the financial and capacity management disciplines of the company. To be fully effective, PPM is best extended to include those initiatives generated from operations.

The extent of PPM is scalable and is to be tailored to the organization's environment. Not all programs/projects need to be managed by a PPM, and thresholds can be set such as below a certain budget limit or less than a certain schedule duration. Similarly, operational managers may consider that certain routine support and/or maintenance projects are best managed outside of a PPM environment.

A key benefit of Project Portfolio Management is that it provides executives with a synthetic view of how programs/projects contribute to the organization's Strategic Intent and fulfill operational objectives. It also assists executives to assess where funding/resources are needed for program/project contenders.

Project Portfolio Management will provide for the periodic review, direction, and allocation of priorities and resources across the portfolio. This will consider:

■ The business unit or organization's strategy and Objectives
■ Changes in the internal or external environment
■ Business operational performance
■ The status, expected benefits, and risks of all portfolio programs/projects

Project Portfolio Management allows for an efficient allocation of internal/external resources and funding for those programs/projects currently in the portfolio or to be included. Each program/project is subsequently assigned a contribution ranking to the portfolio's Strategic Intent.

The PPM is the overarching management and governance process for the identification, evaluation and selection, prioritization, authorization, and performance review of programs/projects within the portfolio, and their alignment to the Strategic Intent and operational goals.

Programs/projects in the portfolio are subordinated to decisions based on their alignment with corporate strategy, viability, portfolio resource availability, priorities, and the evolution of the portfolio contents.

Project Portfolio Management ensures that the portfolio stays aligned with business objectives. This involves following a continuous process by which programs/projects are evaluated, prioritized, selected, and managed at formal points such as a portfolio gate reviews.

1.7.5 Selection and Prioritization of Projects

The assessment, prioritization, and selection of projects are best done by a Project Selection Committee. This will certainly be the case when functioning with a Project Portfolio Management system. Outside of this, informal decision-making should seek inspiration from the Project Portfolio Management system to lend credibility to the process.

The project sponsor should gain an understanding of the organization's formal and informal project selection processes. Being knowledgeable about these processes, providing all pertinent project information to the organization's Project Selection Committee and introducing the proposal to the Committee at the appropriate time will improve the project's chances of being selected.

To implement an effective project selection and priority process, a Project Selection Funnel approach must be instituted and implemented (Figure 1.20).

The Project Selection Committee's roles and responsibilities should be clear:

- Evaluating project proposals based on the selection criteria
- Accepting or rejecting proposals
- Publishing the score of each proposal and ensuring the process is open and transparent
- Balancing the portfolio of projects for the organization
- Evaluating the progress of the projects in the portfolio
- Reassessing organizational goals and priorities if conditions change

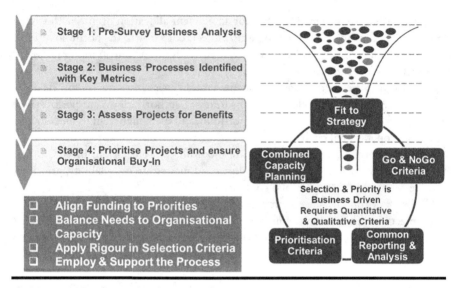

Figure 1.20 Project Selection Funnel.

Initially, the committee can filter out projects that do not at least meet the following:

- Fit in with the organization's strategic plans
- Fit into the existing (or projected) organizational processes
- Compliant with the organization's standards
- Conform to available funding allocations and limits

The contents of the full project proposals assessment criteria will vary between organizations. At a minimum, the Project Selection Committee evaluates how the project proposal:

- Aligns to strategic goals
- Aligns with core competencies
- Responds to stated business issue/problem and goals
- Presents expected outputs and outcomes that are consistent with the goals described
- Demonstrates clearly how the goals will be fulfilled – project plans, schedules, resources
- Includes indicators to monitor implementation progress and performance (intermediary outputs and milestones) toward achieving their final outputs and goals
- Manages risks
- Defines the organizational impacts
- Presents operational sustainability

- Describes clearly and realistically the funding requirements
- Demonstrates benefits and value
- Complies with national or local laws and regulations

Management weights each criterion by its relative contribution and importance to the organization's goals and strategic plan. The Project Selection Committee evaluates each project proposal by its relative contribution or benefit to the selection criteria. The committee assigns a spectrum of values for each criterion ranging from low (0) to high (10). This value represents the proposal's fit to the specific criterion. The aggregate of all assigned values determines the ranking of the project proposal.

The project sponsor follows the selection process, attending the Project Selection Committee meetings as required. The committee determines which projects get approval to proceed to project initiation, which project proposals require more information for further evaluation, and whether some projects should be removed from further consideration. The committee decisions are documented and communicated to the project sponsor.

The main measurement of success for project selection is the consensus of the performing organization management that the proposed projects were weighed fairly and that the ones with a compelling business case received approval.

1.7.6 *Project Portfolio Management System for Project Selection*

An organization will have at a minimum two Project Portfolio Management Systems (PPMS), one at the executive level for transformational change and the other at the operational level for tactical change.

The executive-level Project Portfolio Management Systems will be the overarching approach for the organization and will set the processes and procedures to conduct for all entities.

Each operational unit will institute a PPMS according to its needs, and the core processes will be identical to the executive-level system. However, because of inherent differences across operational units, expanded and tailored PPMS will exist that respond intimately to the nature of the unit's mode of operation.

Thus, an organization will have multiple PPMS, where the transformational change programs and projects PPM at the executive level will hold priority over the operational units' PPMS for tactical projects. Consequently, tactical change projects in operations may need to be on standby, shelved, or abandoned due to constrained and restrained funding and resource availability. Whenever it is deemed that a tactical change project is essential for the operation or not, management may decide to augment the operational unit's yearly budget to cater to the change project.

1.7.7 Geographically Dispersed Projects

Geographically dispersed projects refer to initiatives, programs, and projects that are launched, executed, and implemented across different national and international regions. These are usually the result of organization-wide transformational change programs and projects. Special attention is to be placed on these projects, as these will not only cross national and international boundaries but will also affect the local geography's executive-level and operational unit schemes for change in their respective areas. This topic is further discussed and expanded upon in Chapter 2, "Management of Programs".

1.7.8 Alignment with the Organizational Strategy

As programs and projects are executed, it is imperative that the Project Portfolio Management system ensures that project performance progress is monitored in a timely manner to ascertain that alignment with the established strategy is maintained.

Furthermore, external factors to the organization or internally forced changes, such as evolutional growth or contraction, may well affect the original expectations of the launched initiatives. Additionally, a given project's set of Stakeholders might evolve, leading to different expectations, or the original data on which the strategy was based could change. This is to be coupled with nascent initiatives that will need to be assessed and considered to be included in the project portfolio.

The PPMS must therefore adapt to changes that are to be enacted on the current set of projects in the portfolio and re-establish the ranking and priorities previously set for projects and adjust the plans accordingly, with repercussions to the achievement of organizational objectives, CAPEX and OPEX funding, and resource availability.

The organization, in implementing and utilizing a PPMS, must have an open attitude and a culture of accepting change within a change to achieve organizational goals, by the adjustment of those initially set goals for the benefit of the organization.

1.8 Challenges to Successful Implementation of Organizational Goals

Achieving Organizational Goals is the successful transition of states from old to new, whether the change is transformational or tactical. Organizations must primarily establish a solid foundation for their ability to manage change, by:

- Setting a clear and articulated Vision and Mission
- Defining strategies and Objectives in measurable terms
- Responding to market needs in a timely manner
- Empowering employees
- Focusing on enhancing Return on Investment

The organization must also establish, sustain, and maintain a coherent organic structure and an excellent set of skills capable to manage change projects in response to business issues and opportunities. The organization must also embrace an enterprise-wide culture of Change Management, human resource management accompanied by its infrastructure, and methodology ability and capacity to respond to change.

1.8.1 Agenda for Implementing Strategy

A strategy implementation agenda consists in grouping a comprehensive set of management actions that when combined will bring success to the organization. High on the agenda is for management to underline the compelling reasons for the change and to demonstrate strong leadership and commitment at all levels. This must be coupled with assigning middle management ownership of the change.

The central portion of the agenda focuses on how the change must be accomplished. This requires that a realistic plan be developed with clear Objectives, describing the actions, activities, and tasks that are to be performed. The implementation of the change projects must include ongoing communication with all participating entities and individuals. The agenda must also address Organizational Readiness by dispensing training and education where needed and must include retooling and infrastructural adaptations. Most of all, the agenda must incorporate the necessary funding and provision for expert resources and support for change at the employee level. A governance mechanism must also be included to measure progress toward the goals and to continually adjust the change project's journey.

1.8.2 Strategy-Implementation Tasks

Organizational units must respond by taking actions to implement their allocated contribution to the overall strategic plan and how to successfully achieve it. Multiple tasks shape a manager's action agenda for implementing strategy.

The key components of the strategy-implementation tasks are discussed below.

1.8.2.1 Exercising Strategic Leadership

- Those who hold supervisory/managerial responsibilities must strive and act to gain commitment to the strategy and its accomplishment from all participants of the change initiative.

1.8.2.2 Evolving the Organization's Culture to Fit the Strategy

- Management responsibility must encompass focus and actions to facilitate the units to perform conscientiously and judiciously toward the accomplishment of the strategy.

1.8.2.3 Structuring an Organization Capable of Executing the Strategy

■ Operational units must be structured to enable the strategy to become a reality and act so that the unit's human resources possess the skills and availability essential to successfully execute the strategy. The major tasks to be performed and the decision-making process must also be identified.

1.8.2.4 Instituting Rewards and Incentives

■ The organization must deploy a consistent and transparent performance recognition for employees and operational units that are linked to Objectives and strategy.

1.8.2.5 Establishing and/or Enhancing Coherent Internal Systems

■ Standards, policies and procedures, information systems to provide timely strategic progress monitoring information, and other administrative systems must be employed to enable the organization to assess and improve on its strategy-executing capability.

1.8.3 Obstacles and Pitfalls to Achieving Organizational Goals

Managing change and Achieving Organizational Goals is not a smooth ride. Change will happen in the future in an environment of change. Additionally, change is initiated in a state which might not be completely documented or informed, and unknowns and assumptions abound. The future state envisioned is at best a rational aspiration.

Managing change requires a mentality that accepts change will happen and not in the way that is anticipated.

1.8.3.1 Problems with Successful Implementation

This revolves around how well or badly the existing organization responds to the change and the failure to predict the time and problems inherent in an implementation initiative. This may be a result of:

■ A lack of executive support
■ Strategies poorly forecast or inflexible
■ Not enough focus on people
■ Stifling of innovation

The progress on any implementation will endure internal and external changes that, on the one hand, might not have been sufficiently analyzed and, on the other hand, might be caused by factors totally out of the control of the organization. Unexpected or unwanted changes will affect progress by shifting priorities, which deflect efforts, disable consistency, impact schedules, modify funding, and subsequently cause delays in deploying the initially targeted results.

Without an effective, or existing, Project Portfolio Management System to determine if launched change projects are aligned with strategy, Achieving Organizational Goals will be severely hindered due to misused funding, incomplete and inappropriate resource availability, and wasteful operational activities and commitments that distract attention from the initiative.

1.9 People Are Key

Managing change to achieve organizational goals can only be accomplished by people. Projects will embark all participants in a journey of discovery, as any plan is just a rational speculation of what the planner wants to happen, and not what will happen. People on any project will confront unknowns and uncertainties and will decide on actions to take in the encounter of risks, be they negative or positive.

All employees, from manual to knowledge workers, are of great importance to and in the organization, as their change project participation both physically and intellectually is a fundamental key to success. Creativity and innovation must be allowed to blossom and flourish and should not be stifled by somewhat archaic management styles. The organization must allow individuals to take risks, albeit following rational and logical analysis. Any risk action is directed at events that may potentially occur in the future. Thus, there cannot be any certainty as to the outcome, and the resulting risk action may fail.

Furthermore, failure or success in change projects is not binary. Any calculated decision, any estimate, any plan, and any forecast are approximations of a desired result. The key is to be approximately right within an acceptable range as defined at the outset. Mistakes will happen; however, these have to be positioned in a larger frame.

A management climate where people are afraid of taking risks creates a greater potential for mistakes and failure, as individuals will shun away from extended analysis, creativity, and innovation when confronted with difficult decisions. Rare are individuals who relish mistakes and failure.

The Management of Change by projects culture in an organization must nurture the individual's ability to confront the difficulties as stated above.

- People who are afraid of making mistakes will make mistakes
- People who are not afraid of making mistakes will make mistakes
- The former will hide them and find a multitude of excuses
- The latter will correct them and often will correct them before the mistake is made

As the saying goes, "You learn from your mistakes".

People are the key resources of the organization and change can only be made through people. As a manager, from first-line supervisor to CEO, one major statement must be borne in mind:

Your Goals and the Objectives that you set can only be achieved through the efforts of others.

Chapter 2

Management of Programs

2.1 Chapter Overview

As described in Chapter 1, it is critical for organizations to ensure that transformational and tactical changes are successful and deliver the targeted benefits. This requires two critical factors: first, a well-thought-out strategy that provides the future direction for the organization and responds to its sustainability and growth Objectives. Each organization's strategy for approaching change can be categorized as:

- Reactive – a change in circumstances elicits a response in an immediate or short term
- Proactive – proceeding in a systematic, planned approach and establishing a sequence of activities designed to achieve specific goals and benefits within a period of time
- Continuous improvement – as an ongoing endeavor to gather and analyze data on current and future operational conditions, and the process of enhancing organizational performance accordingly.

The second critical factor is the selection and the successful execution of a strategy and/or tactic to provide the desired change. This is substantially more challenging as the execution of a change as it proves to be significantly more difficult than formulating its strategy.

Organizational change encompasses the innovation of new organizational strategies and a change in how an organization operates and requires a holistic approach

DOI: 10.1201/9781003424567-3

across operational functions through the alignment of people, processes, and technology.

Organizational change will cross functional boundaries; impact both internal and external Stakeholders; and span a longer timeframe when the change is transformational. Thus, change is to be executed with the appropriate business analysis and a comprehensive definition of its scope and Objectives, while affording flexibility to adjust to evolution in and outside the organization's environment.

Irrespective of the type of change to be realized, using formalized Program Management substantially increases the probability of a successful change transformation. Programs related to organizational transformations necessitate a consistent approach to effectively deliver results.

A program consists of related subprograms, projects, and other associated activities that are managed and coordinated in order to achieve organizational business benefits that may not be reached by managing projects individually.

Programs deliver benefits by generating business value, enhancing organizational capabilities, introducing new products and services, developing new capabilities for the organization, and enabling business change.

The bedrock of Program Management is the discipline of project management principles and requires focus on ensuring that the initiative addresses the strategic Objectives most effectively and that target benefits are realized. Programs will cover a broader stakeholder community and widen their scope of interest to address organizational risks.

Several elements are to be addressed by programs of change, where execution performance is to be continuously aligned to strategy while examining alternative approaches to achieve stated Objectives and benefits. The focus on program outcomes and the realization of benefits is the driving reason for investing in change, and managing stakeholder interests, both internal and external, is fundamental. Associated with the latter is the importance of Organizational Readiness such as to enable the enterprise to operate efficiently in its new environment and achieve the benefits and Objectives. The core characteristic of programs, as medium- and long-term initiatives, is that they be flexible to change as overall conditions change.

2.2 Transcribing Strategies into Programs

Transformational and tactical change needs are generated at both the higher level of management and the operational level. Business analysis and the resulting proposed initiatives contribute to the decision-making process by performing organization management to achieve its Strategic Intent and produce business benefits.

Program origination follows the decision to enact a selected initiative. A Program Manager is not as yet assigned to the project until the formalized initiation phase.

Figure 2.1   From strategies to programs.

Nevertheless, the Program Manager should understand why the project is being launched.

Performing organization management will identify and preassign project sponsorship at this stage (Figure 2.1).

2.2.1 Program Origination

The purpose of program origination is to provide a formal mechanism for recognizing and identifying potential programs within the organization, to evaluate programs proposed for the next cycle, and to reach a decision on programs to be selected. During this phase, a program proposal is developed to create a product or develop a service that can solve a problem or address a need in the performing organization. The proposal's Business Case is reviewed and validated and the choice of the proposed solution is assessed. Foremost, the proposed program is gauged for its consistency with the organization's strategic and operational plans. It is also appraised for the value it will create and depending on whether it is within the investment and budget guidelines.

A formal program origination process is important because it creates a focal point for the assessment and analysis of all proposed initiatives, directs funding and resources to the right programs, and maintains the organization's focus on programs that will deliver the greatest value. Each organization has its own program approval process. However, this process revolves around three major steps:

1. Develop program proposal – the Business Case is established and the initial program boundary and parameters are defined
2. Evaluate program proposals – quantitative financial analysis is performed, such as cost/benefit analysis and ROI, and proposed programs are evaluated against a set of preestablished strategic and operational business criteria
3. Select programs – following agreement on the program's feasibility, this is ranked with other proposed programs, and a decision is formally made regarding the program proposal

To make a meaningful evaluation, the decision-making body must possess sufficient information on the program's Business Case and the viability of its proposed solution. The competing programs must be evaluated and compared using a consistently applied methodology and the selection process must consider the program's fit with the organizational strategic/operational plans.

Once a program is selected, funding and/or further management commitment is required to progress to program initiation. Formal program sponsorship is then formally established.

The program proposal process in the origination phase may actually be part of the total budget cycle, serving as the justification for funding requests. In this case, the program proposal is to include a budget estimate for the total cycle and a funding request to proceed to and perform the program's initiation phase.

2.2.2 Development of Business Cases and Business Benefits

The initiative selected to launch a program must be based on a solid Business Case and demonstrate the business benefits to be realized (see Chapter 3 for details on business Benefits Realization). There is no unique structure of the program Business Case; however, it must include the following:

1. The **executive summary** highlights the key points in the Business Case. It states the business problem by providing a thorough and objective presentation of a business issue that requires a timely solution. The summary must highlight the benefits and the Return on Investment. It must also demonstrate how the Business Case responds to the corporate strategic and/or operational alignment
2. The **business opportunity** describes the incentive for the program that the Business Case will describe and propose. The business opportunity includes a definition, a statement of scope, and the Objectives that the program will help the organization achieve. It will explain how the expected outcome of the program supports and aligns with the organization's strategic or operational intent

3. The **organizational considerations** examine the current organization and highlight the areas where the proposed program will affect the future structure.

4. The **market analysis** examines changes in the business environment, such as clients, competitors, suppliers, industry standards, and legislation

5. The **constraints** are internal and external. They will cover the current and future constraints originating from the schedule, resources, budget, staffing, technical, and other limitations that may affect the success of the program

6. The **alternatives** section analyzes the different choices that can be considered and their respective merits

7. The **assumptions** are events on which the proposed solution of the Business Case is based. Assumptions must be validated for a program to succeed

8. The **benefits and costs** provide a comprehensive assessment of both expected benefits and anticipated costs throughout the total Product Life Cycle. These benefits are to be qualified in terms of their impact on the organization's operational performance

9. The **financial analysis** compares benefits to costs and analyzes the value of a program as an investment. The analysis must be credible on the Return on Investment and Payback Period by identifying the ownership costs required for the program development and the operational costs for the utilization of the end-product or service. Programs of an anticipated duration greater than one year must also consider the Net Present Value and the Internal Rate of Return. The financial analysis may also need to include a cash flow statement

10. The **sensitivity analysis** evaluates the program risks, presents the mitigation and contingency plans, and the proposed solution's risk exposure

11. The **proposed solution** should offer an approach to solving the business issue/problem and demonstrate that the program is both viable and beneficial. The proposed solution must emphasize the merits it has over the alternative solutions that exist. The proposed solution should address only the agenda described by the business problem, and the approach description should tie the solution to the existing business problem in cost/benefit terms

12. **Program Objectives** should be explicitly stated and be aligned with the organization's strategic and operational plans and infrastructure. The program key Stakeholders must be referenced as well as the nature of sponsorship and upper management support and commitment

13. An **Organizational Impact Analysis** should describe the anticipated impact of the program on the organization's people, processes, and technology, especially the Organizational Readiness requirements

14. **Budget/resources estimates** (mostly in order of magnitude) should consider all cost components of the ownership and operations cycle, including support, maintenance, and other recurring costs

15. A high-level program **implementation plan** is to be established spanning its initiation to close-out phases. Any plan at this stage can only be tentative and must be considered as preliminary. The subsequent initiation and planning

phases will refine the plan. The proposed program development plan will include:

- Deliverables schedule
- Phase/stage definitions
- Macro activities
- Workload estimate/breakdown
- Program schedule
- Required resources
- Funding requirements
- Program leadership team, Program Governance team
- Program controls and reporting processes

16. The **recommendations** summarize the main points of a Business Case and offer suggestions on how to proceed with the program.

2.2.3 Enabling Documents

A proposal for a program may be instigated from anywhere in the performing organization; however, a program sponsor must be assigned to ensure that the key proposal documents are established. The sponsor would steer the program through evaluation and selection. A business analysis proposal team is often established to develop the Business Case and proposed solution documents, which describe the product(s) of the program, the benefits to the organization, alignment with the organization's strategic and operational plans, and a high-level estimate of the required funding, resources, and costs.

To make a meaningful evaluation, the decision-making body must possess a thorough program proposal. The key enabling document is the Business Case, which provides detailed information on the program's drivers and the viability of its proposed solution.

2.3 Program Selection

2.3.1 Programs and Program Portfolio Management

Programs contend with each other due to limited and constrained funding and resources. It is essential for any organization to establish a solid and pragmatic process for selecting those programs that bring the most value and reap the highest benefits.

Each organization will choose how best to approach the Management of Programs and the level of discipline it wishes to follow to select programs that will fulfill the strategic and operational goals. A haphazard and/or subjective selection technique often leads to missed opportunities and a waste of funding and resources. An objective and informed program selection process does not guarantee total success; however, it will improve the chances of success.

Whatever selection process is followed, a mechanism must exist that centralizes and brings together all program contenders, to be assessed and compared, and from which the performing organization can determine the priorities to allocate to each program. This mechanism has many names; however the most applicable and efficient is a Program Portfolio.

PMI offers a very comprehensive definition:

"A Program Portfolio is a collection of programs and/or programs and other work that is grouped together to facilitate the effective management of that work to meet strategic business Objectives. The programs or programs (hereafter referred to as "components") of the portfolio may be mutually independent or related. At any given moment, the portfolio represents a "snapshot" of its selected components that both reflect and affect the strategic goals of the organization-that is to say, the portfolio represents the organization's set of active programs, programs, sub-portfolios, and other work at a specific point in time".

Program Portfolio Management (PPM) groups programs/projects so they can be managed as a portfolio. PPM ensures that programs/projects and expenditures are aligned with corporate strategy and operational Objectives.

Program Portfolio Management (PPM) is a framework that translates strategy into programs/projects and aligns these to the financial and capacity management of disciplines within the organization. To be fully effective, PPM is best extended to include those initiatives generated from operations.

The extent of PPM is scalable and is to be tailored to the organization's environment. Not all programs/projects need to be managed by a PPM, and thresholds can be set such as below a certain budget limit or less than a certain schedule duration. Similarly, operational managers may consider that certain routine support and/or maintenance programs are best managed outside of a PPM environment.

A key benefit of PPM is that it provides executives with a synthetic view of how programs/projects contribute to the organization's Strategic Intent and fulfill operational Objectives. It also assists executives to assess where funding/resources are needed for program/project contenders.

Program Portfolio Management will provide for the periodic review, direction and allocation of priorities, funding, and resources across the portfolio. This will consider:

■ The business unit or organization's strategy and Objectives
■ Changes in the internal or external environment
■ Business operational performance
■ The status, expected benefits, and risks of all portfolio programs/projects

PPM allows for the efficient allocation of internal/external resources for those programs/projects currently in the portfolio, or which are to be included. Each program/project is subsequently assigned a contribution ranking to the portfolio's Strategic Intent.

The PPM is the overarching management and governance process for the identification, evaluation and selection, prioritization, authorization, performance review of programs/projects within the portfolio, and their alignment to the Strategic Intent and operational goals. Programs/projects in the portfolio are subordinated to decisions based on their alignment with corporate strategy, viability, portfolio funding and resource availability, priorities, and the evolution of the portfolio contents.

Program Portfolio Management ensures that the portfolio stays aligned with business Objectives. This involves following a continuous process by which programs/projects are evaluated, prioritized, selected, and managed at formal points such as a portfolio gate review.

2.3.2 *Aligning the Program to the Organizational Strategy*

An organization's Strategic Intent can only be accomplished by an efficient and focused approach of management by programs. The Strategic Intent is translated into road maps constituting sets of programs/projects, all requiring funding and allocation of resources. Programs/projects are periodically reviewed to confirm their continued alignment with Objectives and their priority is adjusted depending on their performance.

Program Portfolio Management is a pivotal process for the successful fulfillment of Strategic Intents. Depending on each organization, the visibility given to the portfolio by all levels of executive and operational management will be a key critical success factor. If selection decisions are not made at the portfolio level, by default the program portfolio is the end result of individual program choices made independently with little regard for the impact that one program has on the next.

Members of the portfolio executive group or steering committee, sponsors, and key Stakeholders of all programs/projects in the portfolio are to receive pertinent and up-to-date progress/status information on the portfolio performance. Operational management needs to be informed the Organizational Readiness requirements they have to prepare to fulfill the Business Benefits after the completion/handover of any subset of a program/project.

Upper executive management is to be informed on the investment/funding requirements for the medium and long term. In many cases, this is done by a "rolling" quarterly financial statement covering three to five years, or more depending on the organization.

Upper and executive management must inform the Program Portfolio Management about any anticipated changes in the company's direction, to ensure the alignment of programs/projects to the modified Strategic Intent(s).

2.3.3 *Selection and Prioritization of Programs*

A program selection committee best does the assessment, prioritization, and selection of programs. This will certainly be the case when functioning with a Program Portfolio Management system. Outside of this, informal decision-making should seek inspiration from the PPM system to lend credibility to the process.

The program sponsor should gain an understanding of the organization's formal and informal program selection processes. Being knowledgeable about these processes, providing all pertinent program information to the organization's program selection committee, and introducing the proposal to the committee at the appropriate time will improve the program's chances of being selected.

In order to implement an effective program selection and priority process, the program selection committee's roles and responsibilities should be clear:

- Evaluating program proposals on the basis of the selection criteria
- Accepting or rejecting proposals
- Publishing the score of each proposal and ensuring the process is open and transparent
- Balancing the portfolio of programs for the organization
- Evaluating the progress of the programs in the portfolio
- Reassessing organizational goals and priorities if conditions change.

Initially, the committee can filter out programs that do not at least meet the following:

- Fit in with the organization's strategic plans
- Fit into the existing (or programmed) organizational processes
- Compliant with the organization's standards
- Conform to available funding allocations and limits.

The contents of the full program proposals assessment criteria will vary between organizations. At a minimum, the program selection committee evaluates how the program proposal:

- Aligns to strategic goals
- Aligns to core competencies
- Responds to stated business issue/problem and goals
- Presents expected outputs and outcomes that are consistent with the goals described
- Demonstrates clearly how the goals will be fulfilled – program plans, schedules, resources
- Includes indicators to monitor implementation progress and performance (intermediary outputs and milestones) toward achieving their final outputs and goals

- Manages risks
- Defines the organizational impacts
- Presents operational sustainability
- Describes clearly and realistically the funding requirements
- Demonstrates benefits and value
- Complies with national or local laws and regulations.

Management places weights on each criterion by its relative contribution and importance to the organization's goals and strategic plan. The program selection committee evaluates each program proposal by its relative contribution or benefit to the selection criteria. The committee assigns a spectrum of values for each criterion ranging from low (0) to high (10). This value represents the proposal's fit to the specific criterion. The aggregate of all assigned values determines the ranking of the program proposal.

The program sponsor follows the selection process, attending the program selection committee meetings as required. The committee determines which programs get approval to proceed to program initiation, which program proposals require more information for further evaluation, and whether some programs should be removed from further consideration. Committee decisions are documented and communicated to the program sponsor.

The main measurement of the success of program selection is the consensus of the performing organization management that the proposed programs were weighed fairly and that the ones with a compelling Business Case received approval.

2.3.4 Quantitative and Qualitative Methods

Program proposals have to demonstrate viable and realistic financial and economic data for the total costs of ownership and operations. This is applicable only for programs that produce a quantitative tangible benefit that can be measured in monetary terms. Many programs which enhance operational efficiency, increase organizational skills, increase customer satisfaction, boost the organization's credibility, and contribute indirectly to the benefits of other programs will be hard put to calculate the financial effects. Program Managers must be conversant with the following key quantitative techniques:

- Benefit–cost ratio – BCR
- Internal Rate of Return – IRR
- Payback Period – BP
- Return on Investment – ROI

The backdrop to the above techniques is illustrated in Figure 2.2.

A major consideration in BCR, and the other techniques presented below is the notion of the time value of money. Present Value and Net Present Value are used to calculate this.

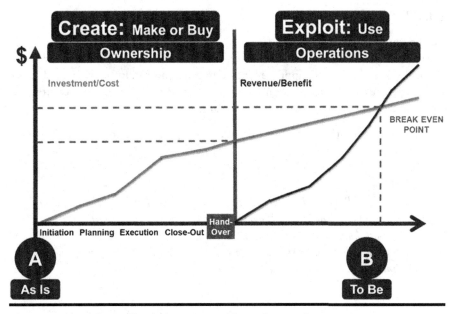

Figure 2.2 Product Life Cycle – ownership and operations.

2.3.4.1 Present Value

Present Value is the value on a given date of a future payment or a series of future payments, discounted to reflect the time value of money and other factors such as investment risk. Present Value calculations are widely used in business and economics to provide a means to compare cash flows at different times on a meaningful "like to like" basis.

The most commonly applied model of the time value of money is compound interest. To an entity that can lend or borrow for t years at an interest rate i per year (where an interest of 5 percent is expressed fully as 0.05), the Present Value PV of the receiving C monetary units t years in the future is:

$$C_t = C\left(1+i\right)^{-t} = C / \left(1+i\right)^t$$

$$PV = C / \left(1+i\right)^t$$

Example:

If the current value C is €1m, the interest rate i is "10 percent" per year, expressed as 0.10, and t is 3 years,

PV = €1m / (1 + 0.10)³

PV = €1m / (1.331)

PV = €751,315

2.3.4.2 Net Present Value

Net Present Value (NPV) is the total Present Value (PV) of a time series of cash flows (also called discounted cash flow or DCF). It is a standard method for using the time value of money to appraise long-term programs. All future estimated input and output cash flows are discounted to give their Present Values. It measures the excess or shortfall of cash flows in Present Value terms.

$$NPV = PV \ input \ cash \ flow - PV \ output \ cash \ flow$$

Example:
Program A and Program B are competing for approval. Program A shows a benefit of 14,000 and a cost of 9,000. Program B shows a benefit of 15,000 and a cost of 10,000. Both have a +5,000 net result before applying PV.
The table below shows that on applying PV and determining NPV over the same time period of 5 years, Program B is more profitable.

Example:

Interest 10%	0.10					
$(1 + i)^t$	1.10000	1.21000	1.33100	1.46410	1.61051	1.77156
Program A	**Year 1**	**Year 2**	**Year 3**	**Year 4**	**Year 5**	**Total**
Benefits	0	2,000.00	3,000.00	4,000.00	5,000.00	14,000.00
PV Input	*0.00*	*1,652.89*	*2,253.94*	*2,732.05*	*3,104.61*	9,743.50
Costs	−5,000.00	−1,000.00	−1,000.00	−1,000.00	−1,000.00	−9,000.00
PV Output	*−4,545.45*	*−826.45*	*−751.31*	*−683.01*	*−620.92*	*−7,427.15*
NPV	**−4,545.45**	**826.45**	**1,502.63**	**2,049.04**	**2,483.69**	**2,316.35**

Program B	**Year 1**	**Year 2**	**Year 3**	**Year 4**	**Year 5**	**Total**
Benefits	1,000.00	2,000.00	4,000.00	4,000.00	4,000.00	15,000.00
PV Input	909.09	1,652.89	3,005.26	2,732.05	2,483.69	10,782.98
Costs	−2,000.00	−2,000.00	−2,000.00	−2,000.00	−2,000.00	−10,000.00
PV Output	*−1,818.18*	*−1,652.89*	*−1,502.63*	*−1,366.03*	*−1,241.84*	*−7,581.57*
NPV	**−909.09**	**0.00**	**1,502.63**	**1,366.03**	**1,241.84**	**3,201.41**

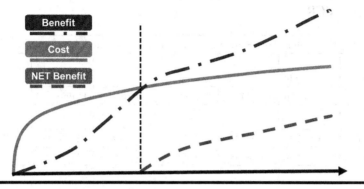

Figure 2.3   Determining benefit–cost ratio.

2.3.4.3 Benefit–Cost Ratio (BCR)

This is the ratio to identify the relationship between the total life cycle cost and benefits of a proposed program. Cost covers both ownership and operational costs, whereas benefits are usually recorded in operations. It is important to define the start point of the cost accrual. The BCR is calculated at a given dateline during operations. This dateline defines at what stage the cumulative costs and benefits can be compared (Figure 2.3).

The BCR is a simple and rapid way to assess the financial viability of a program proposal.

Example 1:
The estimated ownership cost is €1.5m and the estimated operational costs until the chosen dateline is €25.5m, giving a cumulative cost of €27m. The estimated benefits in operations at the dateline are €36m.
*   The BCR is €36m / €27m = 1.33*

Example 2:
The estimated ownership cost is €12m and the estimated operational costs until the chosen dateline is €51m, giving a cumulative cost of €63m. The estimated benefits in operations at the dateline is €49m.
*   The BCR is €63m / €49m = 0.78*

BCR is viable when equal to or greater than one (1). In the program proposal selection process, BCR can be used for ranking.

Example of BCR:

Determine which of the following two programs is the most beneficial for the organization					
Interest 10%	0.10				
$(1 + i)^t$	1.10000	1.21000	1.33100	1.46410	1.61051

Program A	Year 1	Year 2	Year 3	Year 4	Year 5	Total
Benefits	0	60,000.00	300,000.00	400,000.00	450,000.00	1,210,000.00
Costs	−100,000.00	−80,000.00	−200,000.00	−240,000.00	−250,000.00	−870,000.00
Benefits-Costs	−100,000.00	−20,000.00	100,000.00	160,000.00	200,000.00	340,000.00
NPV	0.00	0.00	0.00	0.00	0.00	0.00

Program B	Year 1	Year 2	Year 3	Year 4	Year 5	Total
Benefits	50,000.00	250,000.00	450,000.00	500,000.00	450,000.00	1,700,000.00
Costs	−300,000.00	−200,000.00	−250,000.00	−300,000.00	−300,000.00	−1,350,000.00
Benefits-Costs	−250,000.00	50,000.00	200,000.00	200,000.00	150,000.00	350,000.00
NPV	0.00	0.00	0.00	0.00	0.00	0.00

2.3.4.4 Internal Rate of Return – IRR

The Internal Rate of Return (IRR) is the interest rate that makes the Net Present Value of all cash flows from a particular program equal to zero. The higher a program's Internal Rate of Return, the more desirable it is to undertake the program. As such, IRR can be used to rank several proposed programs an organization is considering. Assuming all other factors are equal among the various programs, the program with the highest IRR would probably be considered the best and undertaken first.

The IRR is determined by numerical iterations. This is cumbersome and using a financial calculator would simplify this process.

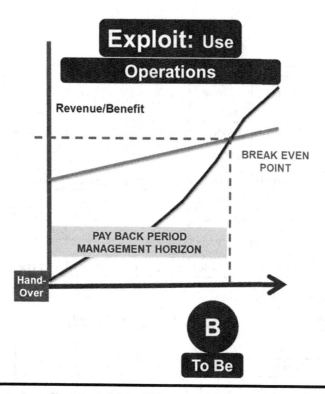

Figure 2.4   Determining Payback Period.

2.3.4.5 Payback Period

The Payback Period in programs refers to the period of time required for the return on an investment to "repay" the sum of the original investment. Graphically, it is the point where both the cumulative cost and benefits curves cross. This point is the break-even point. Shorter Payback Periods are obviously preferable to longer Payback Periods (Figure 2.4).

Payback Period as a tool of analysis is often used because it is easy to apply and easy to understand. When used carefully or to compare similar investments for competing program proposals, it can be quite useful. However, it has serious limitations and qualifications for its use, because, even though the time value of money can be adjusted with Present Value, it does not properly account for risk, financing, or other important considerations such as the opportunity cost. An implicit assumption in the use of the Payback Period is that returns to the investment continue after the Payback Period.

In the example below, the Payback Period (break-even point) will occur during Year 3.

Program B	Year 1	Year 2	Year 3	Year 4	Year 5	Total
Benefits	1,000.00	2,000.00	4,000.00	4,000.00	4,000.00	15,000.00
PV Input	909.09	1,652.89	3,005.26	2,732.05	2,483.69	10,782.98
Cumulative Benefits	909.09	2,561.98	5,567.24	8,299.30	10,782.98	
Costs	−2,000.00	−2,000.00	−2,000.00	−2,000.00	−2,000.00	−10,000.00
PV Output	−1,818.18	−1,652.89	−1,502.63	−1,366.03	−1,241.84	−7,581.57
Cumulative Costs	−1,818.18	−3,471.07	−4,973.70	−6,339.73	−7,581.57	
NPV	−909.09	0.00	1,502.63	1,366.03	1,241.84	3,201.41

Example of Payback Period:

To determine the year of the break-even point for both programs						
Interest 10%	0.10					
$(1 + i)^t$	1.10000	1.21000	1.33100	1.46410	1.61051	
Program A	Year 1	Year 2	Year 3	Year 4	Year 5	Total
Benefits	0	60,000.00	300,000.00	400,000.00	450,000.00	1,210,000.00
Costs	−100,000.00	−80,000.00	−200,000.00	−240,000.00	−250,000.00	−870,000.00
Benefits– Costs	−100,000.00	−20,000.00	100,000.00	160,000.00	200,000.00	340,000.00
NPV	0.00	0.00	0.00	0.00	0.00	0.00

Program B	Year 1	Year 2	Year 3	Year 4	Year 5	Total
Benefits	50,000.00	250,000.00	450,000.00	500,000.00	450,000.00	1,700,000.00
Costs	−300,000.00	−200,000.00	−250,000.00	−300,000.00	−300,000.00	−1,350,000.00
Benefits– Costs	−250,000.00	50,000.00	200,000.00	200,000.00	150,000.00	350,000.00
NPV	0.00	0.00	0.00	0.00	0.00	0.00

2.3.4.6 Return on Investment (ROI)

Return on Investment (ROI) is a performance measure used to evaluate the efficiency of a program investment or to compare the efficiency of a number of different program investments, in a given timeframe. To calculate ROI, first, a timeframe is established, then the cumulative benefit of an investment is divided by the cumulative cost of the investment. The result is expressed as an age or a ratio.

$$ROI = (\text{cumulative benefit} - \text{cumulative cost})/\text{cumulative cost}$$

ROI is a very popular metric because of its versatility and simplicity. That is, if an investment does not have a positive ROI, or if there are other opportunities with a higher ROI, then the investment should not be undertaken.

In the broadest sense, ROI measures the profitability of an investment. As such, there is no one "right" calculation. The calculation can be modified to suit the situation, as it depends on what is included as benefits and costs. A financial analyst and a product marketer may compare two same programs using different parameters. While the analyst may consider the total Product Life Cycle costs, the marketer may only consider the operational costs.

In the example below, Program A has a program development period of one year, and operational use starts at Year 2. The financial analyst will consider costs to accrue from Year 1, while the marketer will start accrual from Year 2.

Program A	Year 1	Year 2	Year 3	Year 4	Year 5	Total
Benefits	0	2,000.00	3,000.00	4,000.00	5,000.00	14,000.00
PV Input	0.00	1,652.89	2,253.94	2,732.05	3,104.61	9,743.50
Costs	−5,000.00	−1,000.00	−1,000.00	−1,000.00	−1,000.00	−9,000.00
PV Output	−4,545.45	−826.45	−751.31	−683.01	−620.92	−7,427.15
NPV	−4,545.45	826.45	1,502.63	2,049.04	2,483.69	2,316.35

2.4 Understanding the Themes of Program Management

Program Management is the key Change Management enabler for an organization. Programs are linked to the Business Case and initiatives that have been selected to achieve organizational goals. Programs will have different Objectives to achieve depending on the nature of the business driver that is at the source of the change.

These drivers can be categorized into the following major types (which can be expanded or renamed by the organization depending on the nature of its business):

- Strategic – transformational change addressing strategic Objectives as identified by upper management
- Operational – tactical change to maintain and sustain current operations as required at the operational levels
- Evolving – organizational change across the enterprise that would benefit from a coordinated program approach
- Client focus – for commercial enterprises, a contractual program in agreement with an external client
- Compliant – program launched to comply with standards and regulations
- Government – program initiated related to a political process or laws

Program Management differs from project management in many aspects, as the table below illustrates.

TOPIC	Program Management	Project Management
Focus on results	Business Benefits	Single objective
Relation to Business Benefits	Drivers for the program Produced throughout the program	Contribute through the production of Deliverables
Extent of the scope of work	Extensive and cross-functional	Specific to objective
Key performance indicators and measurements	Program Mission and Benefits Realization	Schedule, budget, and scope
Number of Deliverables produced	Many and can increase throughout the program	Few, clearly defined within the scope
Governance and monitoring	All-embracing, including organizational functional areas	Specific and bounded
Scope evolution and change	In synchrony with business	Bounded by Objectives
Timeframe, plans, and schedule	High-level and evolving	Specific, detailed, and bounded

The positioning of Program Management is the link and bridge between Benefits Management, which is driven by the organization and is to be aligned to its Business Goals, and project management, subordinated to the program, which produces the Deliverables that enable the organization to exploit the output results.

Program Management provides alignment of strategy and execution, management of Business Benefits, coordination and schedule integration of projects, and assistance to organizational change.

From the approval of the Business Case and subsequent initiative launch, the Business Benefits are confirmed and ownership for their realization is established. As part of the program architecture, a Benefits Map is developed. The map is constructed depicting the link from the program solution, through to the business actions, outcomes, and benefits, fulfilling the organization's overall drivers. The map foremost illustrates a set of organizational enablers to the operational unit, and how they relate to one another, which will allow the achievement of the benefits. Thus, as a result of the program execution the change initiative is implemented. The Benefits Realization is consequently tracked by the performing operational entity.

2.4.1 Boundaries of Business and Program KPIs

As stated, Program Management focuses on enabling the organization to achieve its goals and realize the Business Benefits by building a solution that corresponds to the change needs. The program will build a solution combining project Deliverables to create organizational capabilities to derive Business Benefits. The solutions add or improve capabilities to profit generation, growth expansion, cost reduction, organizational efficiency, and many other goals.

However, Program Management cannot be responsible for the operational performance of the organization and the measurement of the realization of the benefits, as the Business Case is owned by the originating functional entity, and the produced solution is exploited by the operational entity (Figure 2.5).

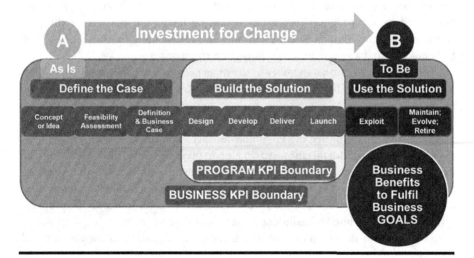

Figure 2.5   KPI boundaries.

2.4.2 *Program Governance*

Program Governance applies an organization's rules and procedures and ensures program procedures are executed as planned and expected by the organization.

A program board will govern Program Managers and, when it exists, the Program Management office. The Program Manager governs Project Managers and non-project out-of-schedule work. The Program Management office, and a change manager where necessary and possible, supports the program and its manager.

Program Governance will encompass comprehensive policies, rules, and procedures to cover key areas such as:

- Program benefit realization and schedule management
- Program Scope Management
- Program funding
- Program quality expectations
- Program risk management
- Program change request management
- Program procurement processes
- Program communication
- Program monitoring, control, and reporting

2.4.3 *The Program Charter*

The Program Charter is a formal document that authorizes the Program Manager to utilize corporate funds and resources to fulfill the program Objectives, or in some cases to only reach the program initiation stage gate. The charter demonstrates management support for the program and authorizes the Program Manager to lead the program and allocate resources as required. The Program Charter is signed and sanctioned by the program sponsor. It states the name and purpose of the program, the Program Manager's name and roles and responsibilities, the identified key Stakeholders, and a statement of support from the sponsor.

The program sponsor, a member of the management team, becomes the highest escalation point and the ultimate decision-maker for the program throughout the program life cycle.

The Program Charter should be distributed to all Stakeholders covered by the program. This will reinforce the Program Manager's authority.

The performing organization management plays an active role following the issue of the Program Charter. Decision-makers must be identified so that they can participate in interviews and document reviews, and at the prescribed program decision points. The performing organization management communicates the program history and background to the Program Manager and the core team, helps define the business origin of the program and its critical success factors and constraints, and commits resources to the program.

The Program Charter should accurately reflect the Vision of the executive management. The program Objectives should be explicit as to how the expected outcome of the program will benefit the organization and help it achieve its Business Goals. Critical success factors (CSFs) should identify the program outcomes that will define the program as a success.

Beyond stating the Program Manager's responsibilities and authority, the key stakeholder's roles and responsibilities, and governance processes, the Program Charter must clearly state the desired outcomes, expected benefits, strategic alignment, and success criteria. The internal and external organizational and environmental constraints should be identified, major program milestones and must-dates defined, and core assumptions described.

The Program Manager will work closely with the program sponsor to identify the core team members and other resources needed to subsequently further define and develop the Program Scope, cost, and schedule. An initial high-level program plan with key milestones is produced and any additional resources that may be needed to progress are identified.

2.4.4 Understanding the Program's Multiple Project Environment

An important characteristic of programs is that they will proceed along a program life cycle, while individual projects within the architecture of the program will proceed along a project life cycle. Although the program life cycle will be the overarching timeframe for the program, projects will be positioned according to their respective sequences within the program and may not exactly coincide with the program's life cycle phases.

2.4.5 Program Management Life Cycle Phases

The Program Charter initiates the start of the program life cycle phases (as further described below) and is a formal step in the whole management of the program process.

The program life cycle can be summarized in five distinct phases. Organizations may have a different number of phases and naming conventions. However, all will require comprehensive reviews at the end of each phase – these will be the gates during which assessment is conducted to verify program performance against planned criteria for exit from the completed phase and to ascertain program readiness to proceed to the next phase.

The program sponsor and the Program Manager must meet at the start of the program life cycle to discuss and share a single Vision for the program. They both need to establish solid bilateral communication, which is later extended to core team members as well as to key Stakeholders involved in the program throughout its life cycle.

Suffice for this section that the principle of phases is retained.
The five phases considered are:

- Identification phase – defines the program's expected benefits and agreement on the program purpose and Objectives. This phase is driven by the contents of the Business Case documentation and benefits to be realized and includes the defined program's strategic goals and Objectives. It is at this phase that the Program Charter is established and a Program Manager is assigned. This phase is a key step as the Program Manager may have had little or no input during the origination phase, as it is the opportunity for the Program Manager to review all the enabling documents provided by the origination phase and to appreciate and understand the quantitative and qualitative methods that senior management have used to select the program

- Initiation phase – the purpose of the program initiation phase is to verify the assumptions and suppositions made in the program origination that led to program selection and approval. During this phase of the program, the overall program parameters are reviewed and refined and key documents are developed. The initiation phase may need to complete the needs assessment or a feasibility study or perform other analysis that was itself separately initiated. In this phase, the program's detailed Business Case and technical specification are developed, as well as the Organizational Readiness operational procedures. This phase will depend on the successful gate review and assessment conducted at the end of the identification phase. The program architecture is established and includes the high-level program Deliverables road map aligned to the Benefits Realization map and plan. Organizational Readiness transition needs are described in conjunction with the affected functional units

- Planning phase – establishes the detailed program architecture of projects to perform, and produces a detailed program road map by allocating the key project Deliverables to projects. This phase will depend on the successful gate review and assessment conducted at the end of the initiation phase. Project Managers develop detailed plans and establish individual schedules for the production of the project deliverable(s). A program-level master plan is produced consolidating and aggregating the major project details for funding, resources, risks, and schedule. Organizational Readiness requirements are enhanced and the program plan is approved

- Benefits delivery phase – describes the delivery of organizational enabling capabilities through the program's constituent projects' Deliverables, including Organizational Readiness. This phase will depend on the successful gate review and assessment conducted at the end of the initiation phase. This phase will cover all individual life cycles of the project, which due to the program architecture will be performed during different periods of the Program Master Schedule. This phase will focus on capturing project progress to determine overall program performance to enable corrections where needed. Produced project Deliverables are reviewed for consistency with requirements, along with the readiness of the organization to effectively exploit these results prior to their introduction in the operational environment. Management reporting on the program will follow the established governance process, and adjustments are made to maintain program alignment with the Strategic Intent and benefit realization goals

- Closure phase – describes the program-level controlled closedown assessment of the Benefits Realization and evaluation of the success of the transition to Operational Benefits. This phase commences when all the project Deliverables have been successfully deployed in the operational environment. Operational Enablers and Business Benefits realized in the organization are captured to complete a program evaluation which concludes with an end-of-program sign-off approval

It is important to consider that different programs will have different timeframes and architecture, and that the above phases of planning and benefits delivery will, in many cases, overlap.

2.5 Program Planning

Program planning is a continuous exercise throughout the program life cycle, from the initiation phase to the closure phase. Planning for the program is at two levels: the program level itself, which starts with the architecture to determine its project constituents and is continuously reviewed during project planning and execution, and at the project level, where detailed plans are constructed and subsequently aggregated into the program master plan, and report on their execution progress, status, and forecasts.

As projects are not initiated at the same time, due to the architecture and the sequence of work across projects, this will require the program planning to be flexible and adapt to the evolution of both the projects' initial plans and their execution performance. Program planning also encompasses the capture of any evolutions in the organization, be it driven by internal or external forces, that modify or change the original contents of the Business Case and Benefits Realization Objectives.

2.5.1 Program Strategy Breakdown

The program architecture is established at the initiation stage. The program co-habits with other programs issued from the portfolio management program strategy framework and has been ranked as a priority for the allocation of funds and resources.

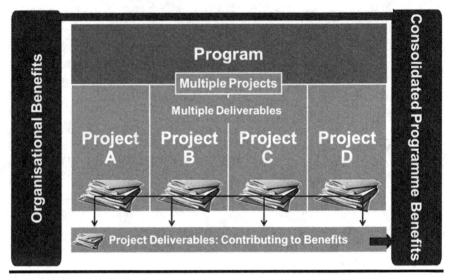

Figure 2.6 Multiple projects architecture.

At the architecture stage, abstraction is made of the availability of resources for the program, as no details at the project level have as yet been developed.

First, the architecture focuses on determining how to produce the consolidated benefits, by identifying the various Deliverables to be produced by the eventual constituent projects (Figure 2.6).

From the identified Deliverables and the assistance of assigned key team members, as identified in the project charter, the program architecture of projects is constructed. This will also identify non-project work which is outside of the project-related tasks.

Non-project work is intended to cover any non-project tasks that require a team member's time and would cover activities such as internal training, general organizational meetings, and the like.

2.5.2 Establishing Program Management Team Structure

The core program team members would have been identified and assigned as approved in the Program Charter. As the architecture develops, individual projects are identified requiring an assigned Project Manager, as well as the need for skills and competencies held by subject experts (Figure 2.7).

Depending on the advanced nature or the lack of details in the Business Case and other enabling documents, business analysts may be required as well as functional managers who will provide more Program Scope contents.

The Program Manager will have oversight covering both program team members and individuals assigned from the operational functional entities. Roles and responsibilities are then established for all program participants.

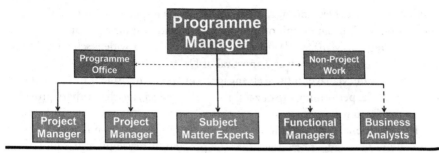

Figure 2.7   Core program team structure.

2.5.3 Roles: Program Manager

The principal role of the Program Manager is to have a constant and direct relationship with business and operations management. This will ensure that the program reports on and remains aligned with the achievement of the Business Benefits goals and Objectives.

Within the program, the central responsibilities of the Program Manager revolve around the definition of the Business Case/scope for each identified project and setting the Deliverables to be produced by each of the projects. The Program Manager, assisted by the core team Project Managers, establishes high-level schedules for the projects, as subordinate to the program's high-level schedule, and sets the governance, policies, standards, and procedures to be followed by and adhered to by the Project Managers.

The Program Manager will initiate and end constituent projects according to the program's high-level schedule, and monitor and control the overall progress of the program through the collection of project performance reports.

The Program Manager is the escalating point for the Project Managers to address performance issues, such as resource, funding, or risk conflicts.

Throughout the initiation phase, the Program Manager will focus on establishing key relationships and engaging decision-makers and influencers.

The Program Manager must go beyond just obtaining the approval of the Program Charter and establish an open and bilateral communication channel with the program sponsor, who is the ultimate decision-maker for the program. Both sponsor and Program Managers must agree on the most appropriate scheme to secure spending authority, to approve major Deliverables, and to sign off on approvals to proceed to each succeeding program phase. The sponsor must also agree to provide support for the Program Manager and assistance in the allocation of resources and funding for the program when the Program Manager encounters difficulties.

The Program Manager must establish the means to capture and maintain the expectations of key Stakeholders who represent the business units that identified the need for the product or service that the program is to deliver. The relationship extends to the decision-makers who have been designated to make program decisions on behalf of major business units that will use, or will be affected by, the product or service. Decision-makers are subsequently responsible for obtaining consensus from their business unit on program issues and outputs and communicating it to the Program Manager. The Program Manager will provide a schedule of program meetings for the review and approval of Deliverables.

The Program Manager must seek engagement and commitment from functional managers of the performing organization. These managers are the primary providers of resources who will perform on the program to produce assigned Deliverables.

The operational organization must be included in the Program Manager's radar, as it will use the product or service that the program is developing. Because of the new product or service, they will be affected by the change in work practices, by modified workflows or logistics, by the quantity or quality of newly available information, and by other similar operational impacts.

2.5.4 Roles – Project Manager

The major role of the project is to deliver the related products/services according to its defined scope and quality standards.

The assigned Project Manager performs the project's subset Business Case/scope derived from the program's Business Case and plans the project according to the inter-project interfaces and program dependencies.

The Project Manager is responsible for the execution and delivery of the project and manages the project risks according to the policies, standards, and procedures set by the Program Manager.

Issues that the Project Manager cannot resolve within the area of responsibility and decision-making scope accorded to them are escalated to the Program Manager.

2.6 Program Management Plan

Establishing the Program Management Plan consists in developing the Benefits Management Plan, the Organizational Readiness processes, and the transition plan of the Deliverables to the functional performing entities.

The Program Management Plan must also describe the life cycle governance and support plans and include the following domains:

- Integration management
- Scope management
- Financial management
- Schedule management
- Resource management
- Risk management
- Communications management
- Procurement management
- Quality management

The organization's approved and used project management processes, tools, and techniques are to be consistently used across the program and the constituent projects.

2.6.1 Program Scope Management

The Program Scope corresponds to the business needs, linking specific Deliverables to the Business Benefits and fulfilling Stakeholders' expectations.

The word "scope" used without any qualification is often confusing as it fails to describe the contents and the state of the program context. For example, at the start of a program, the scope is conceptual and generalized. As the program architecture is structured and projects are identified, the Program Scope contents can be further detailed. However, it is only when the scopes of the constituent projects are detailed that the Program Scope contents can be construed, allowing for changes and evolutions to the originally defined Program Scope.

The word "scope" can be qualified as follows to describe its level of detail and evolution:

- Scope *statement* – describes the major Objectives and describes the program Deliverables
- Scope *description* – describes the needs, main theme, and key components of the program
- Scope *definition* – breakdown of the program corresponding to the functional requirements and its constituent projects
- Scope *specification* – comprehensive architectural design of the program to be delivered containing the individual project scope contents
- Scope *of work* – detailed definition of work to be performed under a contract or subcontracted for the completion of the program. Also called "statement of work" (SOW)

Scope management consists of the processes required to ensure that the program includes all the work required, and only the work required to complete the program successfully. Program Scope management is primarily concerned with defining and controlling what is included in the program and what is excluded. According to the program architecture, the Program Scope is distributed to the corresponding projects as program subsets. And each project will be assigned its goals, Objectives, and Deliverables (Figure 2.8).

The program's scope, derived from the Business Case, is initially at a generalized level. Key milestones related to the program's Deliverables are set in the provisional program timeline, and how these link to the benefits to be enabled in the performing organization. As the projects are initiated and analyzed and planned, multiple iterations will be conducted to align the detailed project scope and schedule contents with the original and provisional program timeline. It is to be expected that

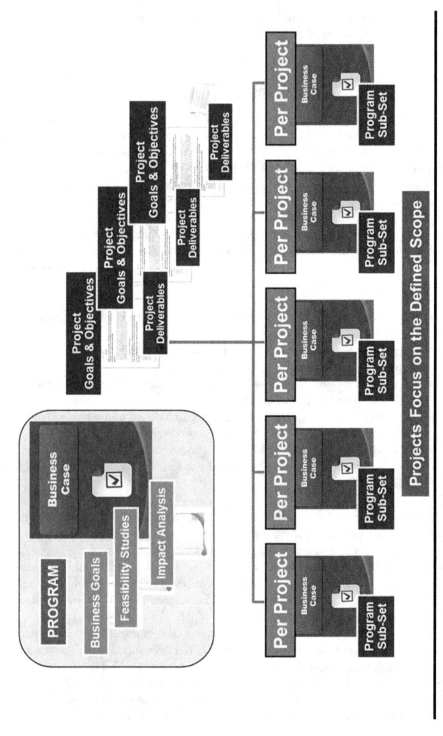

Figure 2.8 Program Scope breakdown to projects.

this synchronization and alignment will challenge originally established timelines, milestones, and eventually, the Program Scope and the corresponding program Business Case.

From the high-level program architecture, the program team can further refine the program's scope. The Program Scope Management process includes the specification of scope, the management of scope changes, and the verification of scope. This process also applies to each project in the program architecture. Thus, scope management is conducted at both the program and the project levels during the planning phase(s) following multiple iterations, and scope Change Management is performed during the execution phases emanating from program-level evolutions and project-level progress.

2.6.2 Key Program Documents

Three major documents are produced for the program: needs assessment, functional requirements, and solution design.

2.6.2.1 Program Needs Assessment

The needs assessment document is derived from the stakeholder needs which motivate the development of the program solution and is to be in direct relationship with the program's Business Case. Needs are often vague, ill-formulated, or not stated. The program needs assessment process clarifies these needs, in a clear and complete manner, and is a major document containing the Stakeholders' expectations. Once projects are identified and launched, their individual needs assessment will be performed and the results matched to the program's needs to determine the discrepancies that may exist.

Generally, it is not possible to meet all the needs within the time and budget available for the program, as the project details of its scope contents, budget, and schedule may prove to be in conflict with the program goals stated in the Business Case. The Program Manager will be required to conduct tradeoffs and prioritization of needs with Stakeholders.

2.6.2.2 Program Functional Requirements

The program functional requirements document elaborates on the needs assessment defining all the major system functions and describes what the product(s)/service(s) (system) will perform, how well it will enable the operational environment, and under what conditions. The subsets of the program functional requirements document identify the Deliverables for each project in the program architecture. Each requirement is linked and traceable to a stakeholder-specified need and is written in a concise, verifiable, clear, feasible, necessary, and unambiguous manner. Each requirement will stipulate the method that will be used to verify its fulfillment.

2.6.2.3 Program Solution Design

The program solution design document defines the architecture, project components, and interfaces for the system to satisfy the specified program functional requirements. The document will take many different forms and vary between organizations. The program solution design is the last major product document and triggers the completion of the program's high-level plan.

2.6.2.4 Impact on the Program Scope

From the key program documents and as a consequence of the development of the needs at each of the project levels, and as the requirements and design documents are elaborated, the overall Program Scope will increase in detail and volume. Stage gates and approvals are to be placed at appropriate milestones in the program and project between the needs and requirements, the requirements and design, and once the design is completed. This allows for alignment with the original program intents and incorporates any required internal or external scope adaptations. The stage gates are also opportunities to review the previously allocated priority to the program and may lead to a suspension or premature termination of the program.

2.6.3 Management of Program Scope Changes

Program Scope Change Management is performed continuously both during the planning and execution phases of the program and during their constituent projects. It is to be anticipated that many Program Scope changes will be captured in the course of project planning, as more details are attained and will require to be assessed for their impact. These impacts may well challenge the original premises of the program's scope contents.

During the project execution phases and the performance data collected, the program scope Change Management will focus on monitoring the progress and status of the ongoing projects and capture and assess changes and how these impact the scope contents of the program and all other projects.

2.6.4 Verification of Program Scope

Program Scope verification is the process of obtaining formal acceptance of the Program Scope by the Stakeholders. Similarly, as project Deliverables are produced, project scope verification must follow the same process, by formalizing acceptance of the completed project Deliverables.

Thus, Program Scope verification is the result of individual project scope verifications to be finalized at the end of the program itself.

2.7 Stakeholder Management

Please refer to Chapter 4, "Stakeholder Management and Engagement", for more details on this topic.

Stakeholder Management is a key skill for all Program Managers.

Program Managers are accountable for the end-to-end management of their programs that enable the organization to realize the desired Business Benefits and achieve the associated organizational goals. The scope of a given program and the results produced involve Stakeholders from both the internal organization and external enterprises. The Program Manager will relate to internal entities and individuals covering a wide range of organizational functions and management levels, while interfacing with providers, suppliers, and public enterprise bodies.

The program's Business Case will identify the Stakeholders that the program will include. The Program Manager should make an explicit effort to understand the full extent of the program and cast wide to capture all potential Stakeholders who have some interest or level of influence that can impact the program.

The collection of Stakeholders will be detailed at both the program and the project levels as per the program architecture. The focus is on the program expectations, which Stakeholders will benefit upon successful program completion, and which Stakeholders could be in conflict with the program's interests. Following the consolidation of the individual project stakeholder maps, the Program Manager will be able to determine which Stakeholders might have conflicting and contradictory interests if and when they are present in more than one project within the program architecture (Figure 2.9).

As will be described in Chapter 4, Stakeholders can be grouped into drivers, doers, and deliverers. The key Stakeholders who are behind the incentive of the program and who have the highest expectations as to the strategic and/or Operational Benefits and business value are in the "drivers" category. The business drivers Stakeholders are the ones holding a direct interest in a given program, as they have a strategic view of the endeavor, express their expectations as Business Benefits to be realized, and expect the delivery of value to the organization from the operational use of the program's results.

The "deliverers" Stakeholders are principally the operational level managers, who will subsequently be exploiting the program's Deliverables and operating in the changed/new environment to achieve the Business Benefits and organizational Objectives.

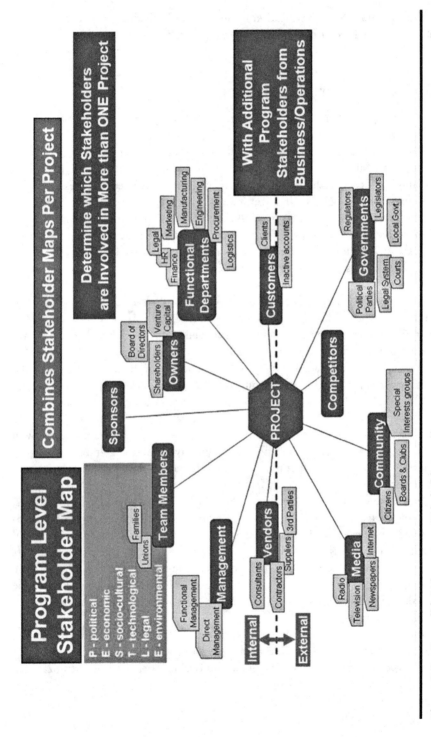

Figure 2.9 Consolidated program stakeholder map.

2.7.1 Stakeholder Management Interfaces during the Program Life Cycle Phases

Program Managers must give consideration to the appropriate involvement and management of Stakeholders. For the Stakeholders' contributions to a program to be successful, a number of factors are to be considered, such as the stakeholder's position and authority within the organization, the degree to which the program depends upon the particular stakeholder to provide a product or service, the level of "social" influence of the individual, and the degree to which the individual is familiar with specific aspects of the business. Each of these factors can be analyzed in more detail during Stakeholder Analysis. In certain cases, for some Stakeholders, their interests will often go beyond the visible program's goals.

For transitional change programs that are extensive across many levels of the organization, and may be spread across many geographies, the Program Manager will deal with a large number of internal Stakeholders and must consider the multiple tiers of stakeholder involvement.

Throughout the program life cycle phases, the Program Manager will focus on achieving the Business Benefits as these are the key project expectations, while matching conflicting expectations between stakeholder "drivers" and "deliverers" and striving to align all Stakeholders to common expectations. The following actions during the program life cycle phases will be conducted:

■ Identification phase – identification and creation of Stakeholders' map
■ Initiation phase – stakeholder expectations assessment, stakeholder power/interest analysis, and program Business Case support evaluation from key Stakeholders
■ Planning phase – Benefits Realization Plan creation from key stakeholder inputs and Stakeholder Management plan development
■ Benefits delivery phase – expectations management, program status and Benefits Realization reporting, and stakeholder support for organizational change
■ Closure phase – status validation of Benefits Realization with Stakeholders

The Program Manager should plan strategies for approaching and involving each individual or group of Stakeholders. Special attention should be given to reluctant Stakeholders, while a specific monitoring mechanism should be instituted for those Stakeholders who may change their level of involvement as the program is performed. When a stakeholder is a group rather than an individual, the Program Manager will need to decide whether all members of the group participate or only selected representatives of the group.

The Program Manager must conduct frequent and comprehensive communications with Stakeholders to enhance the program's success and to provide and receive any evolutions in the established stakeholder expectations. A Stakeholder Communication Plan must illustrate how the information needs of all program team members and other Stakeholders will be satisfied and verified with a feedback loop.

2.7.2 Pragmatism with Stakeholders

The Program Manager cannot expect to align all Stakeholders all the time. The program's stakeholder community changes as Stakeholders evolve within the organization or leave it, and the external stakeholder entities will act and react to socioeconomic and geopolitical changes. Consequently, additional Stakeholders have to be included, while previously identified Stakeholders might be either replaced or no longer active in the program. As programs consisting of multiple projects will usually have a mid- or long-term durations, Stakeholders will be subjected to evaluating their original expectations and request changes leading to additions or reductions to the Program Scope contents and Deliverables. As the program proceeds through its life cycle phases, different Stakeholders might have more or less of an impact on the realization of the Business Benefits and the achievement of the organization's goals.

Due to the dynamics of the stakeholder community and the consequent evolving situation either in their number or in their expectations, the Program Manager must plan for and prepare the program team to review and repeat the Stakeholder Management process in its totality, or in part, several times as the program progresses through its life cycle.

2.8 Program Benefits Map

See Chapter 3, "Benefits Realization Management", for more details on this topic.

For a program to have any prospect of success, it is vital that both requirements and benefits be realistic, clearly articulated, understood by all Stakeholders, accepted and signed off as viable, and supported by a rigorous Change Management process.

The Benefits Realization Plan and map consist of the identification of the expected benefits to be realized in the operational performing entity. Operations have the ownership of Benefits Realization and measurement methodologies are to be instituted by the corresponding unit(s) for achieving the benefits and reaching the declared Objectives.

The Benefits Realization Plan establishes the program's Deliverables production schedule for the enabling of the benefits to be realized in operations and the interfaces to the operational entity and the Organizational Readiness procedures to institute (Figure 2.10).

The Benefits Map illustrates the Objectives to be achieved, the links of the benefits to realizing the Objectives, the positioning of the enablers to the benefits, and the program Deliverables that drive the map.

Figure 2.11 shows an example of a Benefits Map.

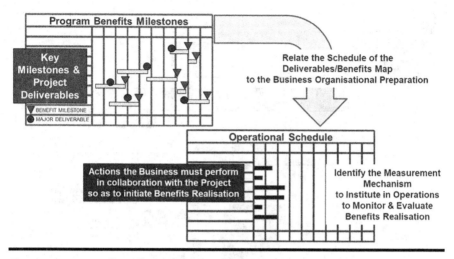

Figure 2.10   Benefits milestones to operations.

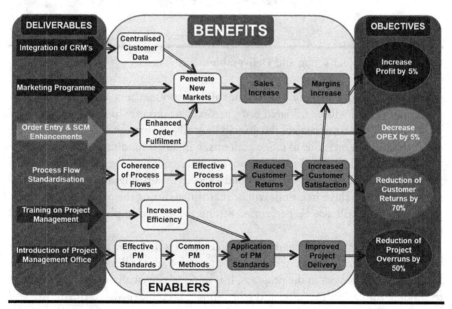

Figure 2.11   Example of a Benefits Map.

2.9 Program Merging of Projects

As described above, the program architecture determines the projects, and the Deliverables for each, that will combine to deliver the benefits enablers to the operational environment (Figure 2.12).

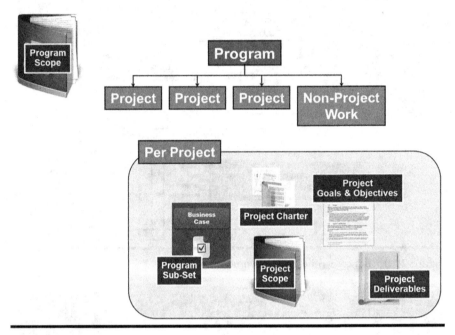

Figure 2.12 Project scope and Deliverables.

Due to the sequencing of projects, not all will commence at the start of the program, and several will be initiated once predecessor projects have been completed. Thus, different levels of project details will be available throughout the program life cycle, resulting in revisions of scope contents, schedules, funding, and resources.

The overarching program plan and schedule combine the individual project plans, and several iterations will be conducted to ensure alignment with the program Objectives. These iterations will be performed during the program and project plan phases and will also be performed during the execution phases of the projects (Figure 2.13).

Tools and templates to manage multiple projects are to be made available that will ensure standard project management methodologies and software applications are used by each project in the program. It is imperative that all projects use the same techniques for project planning and scheduling, Stakeholder Management, risk and change order management, procurement and quality management, and reporting processes.

A supporting PMO can be assigned the responsibility to provide these and monitor their usage.

2.9.1 Consolidating Project Plans and Schedules

As projects are identified, the respective Project Managers develop the corresponding plans, which are then consolidated into the program master plan (Figure 2.14).

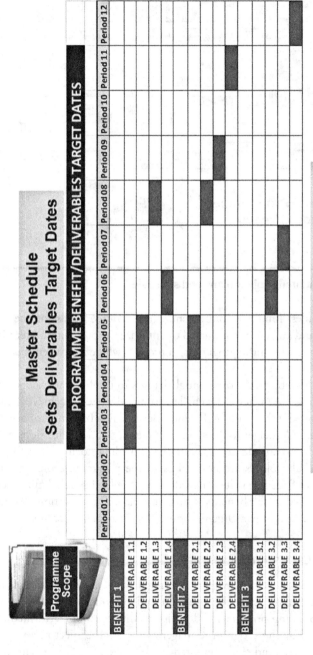

Figure 2.13 Consolidated Program Master Schedule.

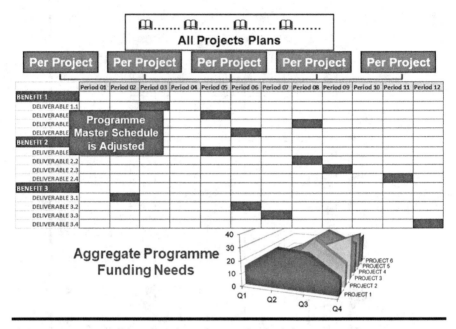

Figure 2.14   Consolidated project plans and schedules.

Combined schedules, funding requirements, and resource needs for the program are established and analysis is conducted to optimize the program master plan, with adaptations made to both the program and projects where and when these are necessary.

Special attention is to be given to potential changes to the Program Scope contents and initial Business Case when these challenge or jeopardize the achievement of the program Objectives and the resulting Business Benefits Realization. Propositions are made by the Program Manager to the sponsorship entity, and the decisions made are then incorporated into an updated version of all enabling documents and program plans, and all Stakeholders are notified as to the program's evolution. The revised Program Master Schedule will set new dates for the Deliverables' milestones, and these will be reflected in the Benefits Realization map. The operational entities will also be required to revise their Organizational Readiness plan schedules.

2.9.1.1 Creating Cross-Project Links

Critical to the elaboration of the Program Master Schedule is establishing links between the projects internally to the program and the interfaces between the program and other ongoing programs and projects. This must be performed at a group meeting chaired by the Program Manager, where each project presents the sets of outputs it produces and the inputs required by the project to perform its activities.

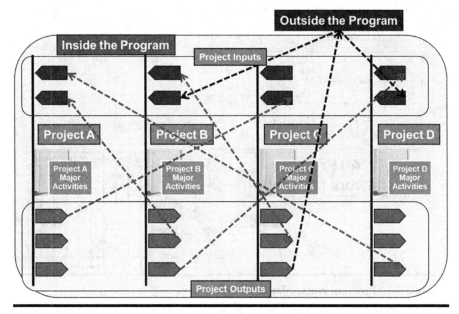

Figure 2.15   Program cross-project links.

The technique to use is comparable to the hierarchical input process output (HIPO), where the processes are the activities to be performed by each project, and the outputs are the internal Deliverables produced by a project to be linked to inputs to another project.

The program team proceeds to link every project input to the corresponding project output. It is imperative that standard and equal naming conventions are used to match outputs to inputs (Figure 2.15).

All inputs to projects must find a corresponding output from either a project within the program or another program outside the boundary of the program. If an input is deemed to originate from within the program and no project produces it, then, after analysis, the required output is added to the corresponding project, and the activity list of that project is enhanced.

Project outputs must find a corresponding input, either to another project within the program or as an interface to another program. If the output has no correspondence, then, after analysis, the output is either added as an input to an internal project, thus enhancing the corresponding project's activity list within the program, or the output is deemed superficial and the generating activities are deleted from the producing project. Care must be taken to ensure that the output is not an interface to another program.

The HIPO technique can be considered to be heavy and cumbersome; however, it will be impossible to produce a Program Master Schedule without establishing the respective links between projects within and outside of the program.

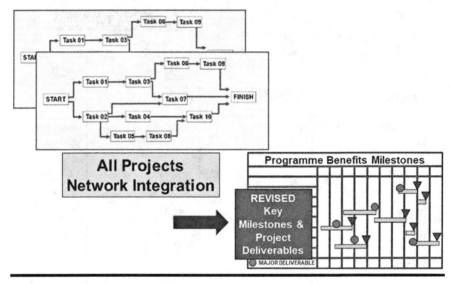

Figure 2.16 Program consolidated activity network.

2.9.1.2 Consolidating Project Networks

The Program Master Schedule is established by merging all project schedules. It is essential that the cross-project links have been determined previously and most importantly that all projects use the same project management software tool. The program benefits milestone plan produced is revised where necessary and the modified dates are adjusted in the corresponding project schedules (Figure 2.16).

2.9.1.3 Aggregating and Leveling Resource Requirements

The Program Manager, along with the respective Project Managers, will analyze the total program resource needs for each type of resource, be it human, materials, equipment, or services/facilities. These are mapped against the organization's capacity plan and providers' availability (Figure 2.17).

2.9.1.4 Consolidation and Resolution of Program Plans

The major challenge Program Managers face results from the consolidation of project plans, during project planning, and subsequently during project execution.

The program and project scopes are adjusted where necessary as detailed project plans are produced and project execution performance is recorded. Additional to the resource plans from each project, as described above, total funding requirements are established for the program life cycle budget. Individual project risk plans are collected and amalgamated and the overall program risks are reviewed and analyzed.

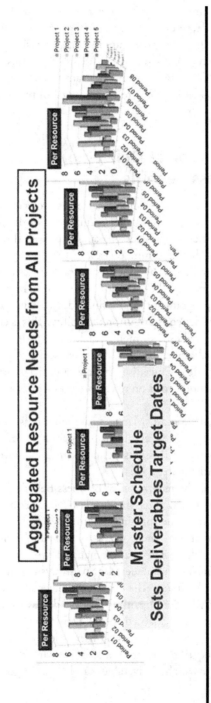

Figure 2.17 Program aggregated resource requirements.

The Program Management team then determines the consequences of the program master plan, schedule, resource needs, funding, and risks, and modifies accordingly. This will involve adapting the program roadmap, revising the Deliverables and benefits milestones, and eventually revisiting the program's Business Case.

All changes to be incorporated into the program must be validated by the sponsorship group, and all Stakeholders are to be informed of the resulting decisions.

2.10 Delivering Project Deliverables and Business Benefits

Program life cycle execution is of a complex nature as projects are performed at different rhythms and schedule timelines. The formal start of the program's execution phase commences with the first project's execution phase. The remaining projects may still be remaining in their planning phases.

The initial complexity lies in the inherent characteristics of capturing real performance data and comparing the variances to the program and project plans. Additionally, when the program interfaces (in and out) with other programs, the overall program plan is under constant unpredictability. This is compounded by any changes in the organizational environment which cause reviews and analysis of the original Business Case and scope contents. The complexity of program execution is further heightened when the organization operates a Project Portfolio Management system, which holds a superior position to the program's priority level, funding, and resource distribution.

In summary, seasoned Program Managers are conscious that the planned and scheduled delivery of project Deliverables will experience variances that are to be addressed so that the goals and Business Benefits can be achieved within a narrow span from the original target dates.

The ultimate impact is the realization of Business Benefits and the accomplishment of the organization's goals. Thus, the Program Manager must constantly relate to and communicate with the performing organization to review existing delivery plans and establish adaptations where necessary. The Program Manager may be assisted in this area by a dedicated change manager who would ensure that the program-to-operations transition pursues clear pathways that link Deliverables and outcomes to Operational Enablers. The change manager would also assist the performing operational entity in the Organizational Readiness requirements and plans that need to be established and adapted to the evolution of the program delivery, and also establish agreed-upon benefits-tracking metrics.

During program execution, progress and status are reviewed with the key Stakeholders to determine which Deliverables can be rearranged to allow for earlier "quick-win" benefits.

2.10.1 Implications of Different Milestones and Deadlines

There are four schedules in a program that are synchronized and require frequent reassessment during program and project execution. The Program Master Schedule is built by the consolidation of the individual project performances. The program Deliverables milestone schedule links directly to the Benefits Map and the schedule dates for the Operational Enablers (Figure 2.18).

As each project is launched in execution, the program and Project Managers conduct a review to ascertain the continuing alignment of the project to the Benefits Map and its synchronization with the program. Evolutions to the project's scope, Objectives, and Deliverables are assessed, as well as modifications to the originally planned project's resource and funding needs and schedule. The program master plan is adapted as needed.

During project execution, actual performance data is collected and the original project plan is reassessed. This would involve again a detailed review of the project's scope and Objectives, organizational constraints, resource and funding needs, the project schedule, and the impact on the Program Master Schedule.

Throughout the program's execution phase, all project schedule performances are to be monitored and changes are to be applied to the program master plan, program Deliverables milestone schedule, and the Benefits Map. If and when necessary, changes are to be made in the prioritization of projects in line with the evolved schedules for the Deliverables and benefits.

The Program Manager, following sponsorship group approval, adapts the Program Master Schedule, as necessary. Changes implemented in these plans will also have to be reflected in the Organizational Readiness plans established by or with the performing operational unit.

2.10.2 Effective Resource Assignments and Task Performance

From the initial consolidation of individual project plans, and the execution progress data collected for those projects in execution, the Program Manager must determine the efficient use of resources across the program and projects, including those from the business and operations entities.

Variations in the project execution delivery schedules will require the Program Manager to reassess the distribution of resources across the program and analyze the consequences of the reassignment or transfer of resources between projects in execution. All current schedules at the program and project levels are to be adjusted to reflect allocation changes.

When these changes are within an acceptable variance range for the program, the manager can apply these adjustments without seeking authorization from the sponsor. Otherwise, no such resource and schedule modifications can be applied without prior approval from the sponsorship group.

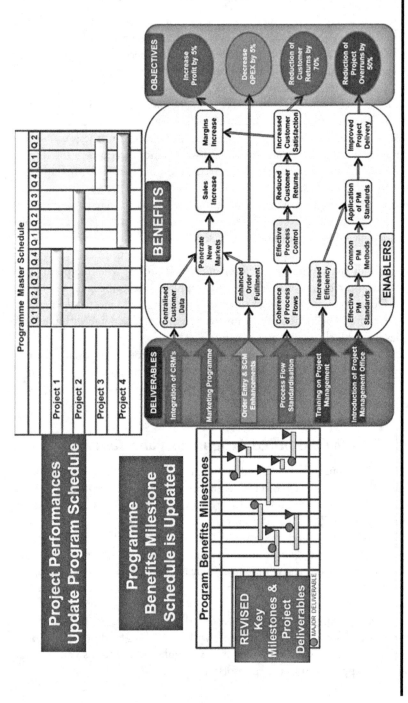

Figure 2.18 Program Master Schedule progress and the Benefits Map.

2.10.3 Managing Change Requests at Program and Project Levels

Program Managers must expect changes throughout the program life cycle.

Substantial effort must be devoted to building consensus from management with a meaningful understanding of the rationale for change requests, agreement and commitment to the nature and consequences of change, and monitoring and refining the transition process.

Change requests must follow a standard process from change request/elicitation, change impact analysis, and change approval (or rejection) to change order implementation. A standard process would consist of the steps mentioned in Figure 2.19.

A Program Change Control Board (PCCB) is a critical structure to be instituted, where all changes to the program are funneled and can initially be filtered to be accepted or rejected. The PCCB will decide to proceed and fund the program and project impact analysis to be performed, and then determine if the change request can be implemented. As a result of a positive decision to pursue and apply the change(s), the Business Case is adapted to the approved changes, the program and project scope contents, funding, resources, and schedules are updated, and the Benefits Realization and Organizational Readiness plans are modified.

When changes originate from key Stakeholders, who are "drivers" of the program, an assessment is conducted to determine the impact severity on the current program status. Change requests that are either within an acceptable range of variance to funding and schedule or mandated by management will follow the Change Request Process and be applied to the program. The impact analysis must define how current ongoing projects will be affected and how projects still waiting to be launched will need to be reviewed. When change requests are outside the current boundaries of the program or are imposed due to changing organizational environment reasons, this may cause upheaval in the contents of the Business Case and all subordinate plans and schedules. The original organizational goals may no longer be attainable and management decisions will be required to ascertain if the targeted Business Benefits are still viable. The Program Manager must demonstrate flexibility to modify and adapt the program, and the subordinated projects, to reflect changes to be made to the existing program and project baselines and scope contents. The operational performing entities must be involved to establish how the Organizational Readiness plans need to be modified.

Change requests may also originate at the project levels. The reasons are varied and will revolve principally around scope contents, technical, financial, resource, risk, and outsourcer/provider motives. The program team's assessment of these changes must consider the impact on the individual project's status as well as analyze the broader impact on other ongoing and future projects. When change impacts are within an acceptable range of variance to funding and schedule, the program team may proceed without authorization from the PCCB. Otherwise, the change request(s) are escalated to the PCCB and the standard process ensues.

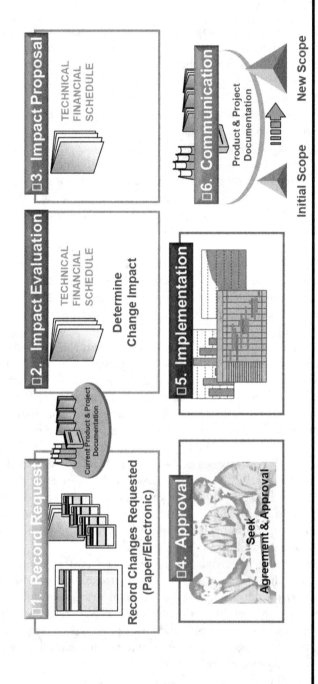

Figure 2.19   Change request process.

2.10.4 *Program Progress Tracking and Evaluation*

Progress at the program level focuses on monitoring that the overall performance is steered to the Benefits Realization. Program Scope Management is to be rigorously applied while organizational changes to the Business Case are assessed. Deliverables milestones are tracked from the project performance levels. The Program Manager and the team will frequently review and redefine program and project plans and schedules according to the progress and forecasts emanating from the status of projects in their execution phase, and how this impacts projects yet to be launched in execution.

Performance-detailed tracking is at the project level and will comprise schedule progress, budget and resource usage, variations, risk responses, and change requests. Project reports highlight information to help aggregate the information at the program level. Some typical areas of concern for Program Management include:

- Project Deliverables meeting scope content requirements
- Effective tracking of benefit enabling
- Scope contents, schedule, and cost management
- Sufficiency in resources availability and funding needs
- Dependable estimates and teams adhering to project schedules
- Appropriate identification and management of risks, issues, and changes

Project tracking will consist of standard monitoring and control activities on status, progress, and forecasting. All collected performance data is mapped against the Program Master Schedule and the Benefits Map schedule. The Program Manager aligns and adjusts these according to the performance progress.

The program risk plans and responses are monitored along with those from the project risk plans and responses. This will enable the Program Manager and the team to measure cross-project impacts and the subsequent variations to the Deliverables and Benefits Realization schedules.

2.10.5 *Ensuring Organizational Readiness*

Organizational Readiness is a major transition preparation for the operational entity to facilitate the acceptance and deployment of the program Deliverables to enable its operation to produce the desired Business Benefits.

An Organizational Readiness Assessment is conducted to determine the extent of the multiple and simultaneous changes required to be made in the operational unit. Preparation is to be planned to cover changes to people, processes, and infrastructure, covering:

- Operational staffing requirements and training of employees for re-skilling
- Institution of new or modified operational procedures and methodologies and workflows
- Operational evolution of property assets, production assets, and technology platforms

Operational management is responsible for the development of the Organizational Readiness and transition plans, with the assistance of the Program Manager. This must involve the pertinent people in the design and implementation of changes and seek to secure buy-in from those involved and affected. The Organizational Readiness transition plan is focused on assisting the people who are affected and impacted by the change to be ready to adapt to the changes by ensuring that they have a comprehensive understanding of its scope. The plan must include schedules for the training of the appropriate operational unit resources on the changes to processes, procedures, methodologies, and systems.

The program team and the operational unit(s) must work in harmony to communicate, inform, and explain to the functional departments the need for change and its goals, purpose, and rationale for achieving the Business Benefits. The transition plan and schedule will detail the implementation and transition, roll-out and ramp-up activities, and the resources to be allocated to these. A Business Benefits measurement mechanism is to be instituted, accompanied by a corresponding support process for raising issues and deriving resolutions.

2.11 Close-Out of Individual Projects and Program

2.11.1 Project Close-Out

The close-out of individual projects in the program is a critical event that determines and signals if the program's Objectives remain on course to be accomplished and will enable the organization to proceed to realize the Business Benefits. The Program Master Schedule, initially elaborated at the start of the program, and further enhanced as projects develop their individual plans and schedule, must be reviewed thoroughly to assess impacts on the program's remaining scope, Deliverables, funding and resources needs, and risks.

Individual projects within the program are closed once their respective Deliverables have been accepted. Acceptance, according to the preestablished criteria, must ensure the project deliverable handover fulfills its current and latest scope contents. This covers:

- Status of completed Deliverables
- Additional Deliverables and/or abandoned Deliverables
- Quality measures of Deliverables
- Outstanding defects
- Client formal approvals
- Contractor deliveries, where applicable

The Deliverables may then be transferred to operations and exploited by the receiving organization. Additionally, other projects within the program that are dependent on the successful closure of a predecessor project can be initiated, either in their planning phase or during execution. It is to be noted that Deliverables produced by a project from outside the domain of the program must also undergo a cross-program acceptance.

The Program Manager then conducts a review of the status of the remaining ongoing projects and those still to be launched in execution and assesses how project Deliverables that interface with other programs synchronize with the existing program master plan schedule. This will cover:

■ Adjusting remaining project schedules
■ Requesting new detailed plans from each Project Manager
■ Reassigning resources correspondingly
■ Ensuring support to the operational entity for the produced Deliverables
■ Communicating to sponsorship groups and Stakeholders

A reorganization of project schedules may be necessary according to the program status with regard to the outstanding activities to perform and their needs in resources and funding. As a consequence of the review, the program master plan schedule may need to be reestablished along with the various ongoing Deliverables milestone, Organizational Readiness, and Benefits Realization schedules. This will require management and sponsorship group approval.

For completed projects, and when team members are no longer required to participate in the program, the project team is demobilized and other resources and infrastructure assets are released. Otherwise, team members can be deployed to other projects within the program.

Of great importance are the recognition of team members' performance and the provision of rewards that may have been budgeted.

2.11.2 Program Progression and Close-Out

Program progression relates to the ongoing production and handover of project Deliverables to operations, according to the Program Master Schedule. During program progression, the Program Manager will conduct the transition and go-live assessments with the operational entity; verify the operational readiness schedule for the go-live with the functional department(s); coordinate the schedule with vendors, contractors, and other providers; provide operational response team to assist the operational roll-out; confirm that operational support systems are in place; provide a disaster recovery business resumption plan; and ensure Operational Benefits Realization metrics have been established.

The program is deemed to be completed when the last deliverable has been produced. The program close-out occurs when all program Deliverables are produced, verified, and validated for compliance. The Program Manager will conduct an assessment with the concerned operational Stakeholders on the exploitation of all the program Deliverables and the operation's ability to enable the achievement of benefits and goals. The assessment will review the use of the delivered results; the workflow and process improvements; operational staff training delivery; and the progression by operations to realize the benefits.

The official and controlled closedown of the program signals the transition of responsibility for benefits monitoring to the operational organization. Meetings are held with the sponsorship group, key Stakeholders, and business managers to review total program results and assess the program's performance against the business and benefits Objectives and evaluate the program performance throughout all phases. A final perspective on the performance of the program/project team is developed and a summary of the lessons learned from the program/project development and execution is recorded.

The governance board finalizes the program close-out and reviews and approves the program's contribution to the desired benefits. The program organization is disbanded and the core program team can then be redeployed. The program infrastructure is released and lessons learned are documented.

According to the organization's instituted mechanism, the Program Manager and the core team members are to be recognized for their contribution and duly rewarded.

2.12 Challenges in Managing Programs

Program Management is the foremost method to enable strategic and tactical changes in an organization. The extent of these changes is broader than that of an individual project as programs facilitate and enable change across an entire organization, within a specific division or a single operational department.

Due to the scope and breadth of programs that have a mid- or long-term timeframe, their contents and target goals will be submitted to changes in an environment of change. The evolution of the originally stated Objectives is to be expected as internal and external forces affecting the organization will be placed upon the program. Beyond these forces and those originating from technical or environmental sources, most program failures are people related. These failures span a range of causes, the principal ones being a lack of strong leadership, organizational commitment, and program sponsorship and support. The inadequacy of Program Management skills leads to underestimating program complexity and a lack of understanding of stakeholder expectations, aggravated by ineffective cross-functional communication. Associated with these factors, program failures are also due to a lack of integrated program planning, poor requirements management, ill-defined success metrics, and a lack of Organizational Change Management. A program is destined to fail when there is insufficient or lack of funding or availability of resources.

The goal of effective Program Management is to define, organize, plan, and execute so that the causes of failure are minimized, as they will never be totally eliminated. However, there is no magical formula for program success. Nevertheless, Program Management can call upon a variety of skills, processes, and methods, from the initial evaluation of the program's Business Case to the measurement of project outcomes and benefits, and be supported by a robust Project Portfolio Management framework and a Project Management Office structure.

The three overarching responsibilities of the Program Manager are effective governance, Stakeholder Management, and Benefits Management.

To reach levels of satisfactory program success, the following key areas are to be fully integrated and understood in the organization's culture and foremost within the Program Management team, sponsorship group, and Stakeholders.

- Understanding Strategic Intents and Business Goals
  - Organizations set Business Goals and establish Strategic Intents in concert with continuous improvement, Business Process Reengineering, and enabling technology
  - Business Benefits arise when investments enable the organization to perform more efficiently and effectively introduce new ways of working and/ or operate in a new way
  - Business Benefits drive initiatives and programs
- Aligning Stakeholders
  - Identification of all internal and external Stakeholders affected by the program
  - Appreciation and recognition of their interests
  - Maintaining stakeholder expectations and involvement throughout the program
  - Establishing Stakeholders' level of participation at all stages of the program life cycle
- Program Scope Management
  - Program Scope content corresponds to stakeholder expectations
  - Scope evolutions and changes are reflected in the program and projects
  - Benefits Realization and Objectives target change in consequence to the evolution of Program Scope contents
- Program leadership and team dynamics with Project Managers
  - The Program Manager's role is strategic and focuses on the Strategic Intent measured by the implementation and accomplishment of a business initiative and the realization of Business Benefits
  - The Program Manager is responsible for a team of Project Managers, whose role is tactical and is centered on the planning of the project, performing its tasks, and producing specific Deliverables to fulfill the project's specifications to schedule and within budget
  - The Program Manager must have both people skills for the leadership and motivation of the team, and project management skills to guide and support the Project Managers

- Program Governance
  - Program Governance must be consistent with organization policy and must enable program decision-making, establish practices to support the program, and maintain program oversight
  - Program Governance scope is to ensure that the goals of the program remain aligned with the strategic Vision and operational capabilities and approve, endorse, and initiate the program and secure funding
  - Well-understood agreements as to how the governance and sponsoring organization will oversee the program must be clearly established
- Resource management and funding across projects
  - The Program Manager addresses resource needs across the projects that constitute the program
  - All internal and external human, material, equipment, and services resources required by the constituent projects are consolidated into the program resource plan
  - The Program Manager must be supported in the program funding needs as required by the projects' planning and executing phases
  - The Program Manager will use negotiating techniques with key internal and external providers to secure resources for both the projects and operational needs
- Consolidated risk management
  - The Program Manager focuses on all project risks as well as the business impacts
  - The program risk plans and responses are communicated to all Stakeholders and their evolution is made transparent to that community
- Consolidated program reporting
  - The Program Manager is solely responsible to collect and review all project reports and present the program summary to the sponsorship group
  - Reporting will encompass program status and forecasts, including program financial statements
  - Program Deliverables produced and the benefits enabled in operations are reported including scope evolutions and recommendations
- Establishing a Dynamic Business Benefits Map
  - The Benefits Map is developed and maintained by the Program Manager and represents how project Deliverables relate to the Business Benefits and Objectives to achieve
  - The Benefits Map is updated in response to the evolution of the program execution and changing business needs
  - Measuring Business Benefits Realization
  - The Program Manager ensures that objective measurement methods are deployed in operations to demonstrate that the organization is enabled to realize the Business Benefits

- Evaluation of Operational Effectiveness
  - The operational entity will progressively ramp up its operation to deliver the Business Benefits
  - The Program Manager, in coordination with the operational unit, institutes the operation's target values and measurement mechanism, the timeframe in operations when improvements are expected to start, and the frequency of measurement reporting

A program delivers a business solution
  The business operation exploits it to deliver the Business Benefits

Chapter 3

Benefits Realization Management

3.1 Chapter Overview

The most important reason for considering Business Benefits is that an organization wants to achieve them. A business benefit is the outcome of an action or decision for change that contributes toward achieving one or more business Objectives. The benefit is considered positive by any or all Stakeholders participating in the change. The benefit may be quantitative or qualitative, as explained further in this chapter.

Funding for the realization of Business Benefits is the key driver for all change initiatives which launch programs and projects. Target Business Benefits are initially identified in the business analysis stage and are further assessed to decide on the necessity to launch an initiative of change. Once the initiative is deemed to be worthy, corresponding programs and projects are subsequently implemented to create the results to be exploited in the operational units. It is only during operational performance that the realization of Business Benefits can be measured to determine if they have been achieved. Thus, Benefits Realization Management extends from the initial business analysis to the effective change in the operating environment.

Business Realization Management, which will call upon the organization's capacity for funding and the resources to perform the change, is intimately related to Achieving Organizational Goals (see Chapter 1), Management of Programs

DOI: 10.1201/9781003424567-4

(see Chapter 2), and its relationship with Stakeholders (see Chapter 4), as all are engaged and perform in unison in Enterprise Project Management, supported by an efficient Project Portfolio Management and an effective PMO (see Chapter 5).

Business Benefits Realization is about ensuring that programs and projects deliver results and enablers for the forecasted benefits as identified in a Business Case or project charter document.

Obtaining a Return on Investment is critical, and Right Projects

Organizations must have the structure, capacity, and skills for their Programs and Projects to be capable of delivering the required Business Benefits. The unequivocal role of a Project Manager is to deliver the project to ensure the scope, schedule, and budget components are accomplished. Furthermore, the assigned Project Manager must possess an entrepreneurial and business-minded awareness to understand and comprehend the rationale driving their organization's investment strategy.

Business Realization Management is holistic and spans a timeframe from inception to finalization. This requires a structured approach to managing change and achieving Objectives. A stand-alone operational performance measurement process divorced from Business Realization Management is pointless.

3.2 An Organization's Challenges

This section briefly reviews the challenges encountered by organizations. Please refer to "Achieving Organizational Goals" in Chapter 1 for full details on this topic.

3.2.1 Organization's Ability to Meet Market Requirements

The major challenge for any organization is to sustain and maintain its operation and pursue a constant need for continuous improvement and innovation. The organization will strive to balance its strategic direction with its operational performance, while addressing situations requiring to be reactive or proactive (Figure 3.1).

Major internal forces are at play simultaneously, all contributing to contesting the level of performance the organization aims to achieve. The core forces requiring particular attention as to their performance are:

■ Analyzing and reviewing the market, the competition, services, and products on offer
■ Delivering end-to-end solutions within time-to-market constraints
■ Transforming key processes to strategic capabilities
■ Redefining organizational structures and empowering people
■ Facilitating enabling technology

Figure 3.1 The organization's competitiveness.

3.2.2 Making Strategic Investments

All the above forces will experience internal changes because of their inherent characteristics, and the organization will be summoned to be proactive, or else face a situation that calls upon it to be reactive.

The organization must also integrate the results of several external influences, such as evolving competition, financial constraints, evolving marketplace, temporal constraints, socio-economic and geopolitical disturbances, impacts of political and regulatory changes, and globalization.

Operating in a stable environment is an ideal situation.

Change is the only certainty in a world of uncertainty.

3.2.3 The Organization's Strategic Intent

Organizations set Business Goals and establish Strategic Intents in concert with continuous improvement, Business Process Reengineering, and enabling technology (Figure 3.2).

The core intents of the organization are meeting its Business Goals as it manages its performance and approaching the way it operates to create value in a proactive manner rather than a reactive one.

The organization will focus on two paths simultaneously: Strategic Transformational and Tactical Operational.

Figure 3.2 Meeting Business Goals.

3.3 Realizing Benefits from Investments

Business Benefits arise when investments enable the organization to perform more efficiently and effectively by introducing improvements to the ways it performs and/or making it operate in a new manner.

Business unit managers and users have the responsibility to generate the Business Benefits by introducing the necessary change initiatives and programs to their organizational plans.

Program management provides an alignment of strategy with the execution of Business Benefits Management, providing coordination and schedule integration to the organizational change plan.

3.3.1 Benefits Management – Program Management Focus

The focus of program management is to deliver a systematic solutions approach for adding or improving capabilities. The focus is on the organizational business unit, and the derived expected benefits and its outcomes.

Programs combine Deliverables to create capabilities and enablers to achieve Business Benefits, whereas projects create individual Deliverables.

Please refer to Chapter 2 for further details on the Management of Programs (Figure 3.3).

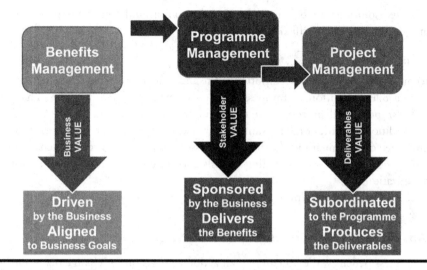

Figure 3.3   Positioning program management and Benefits Management.

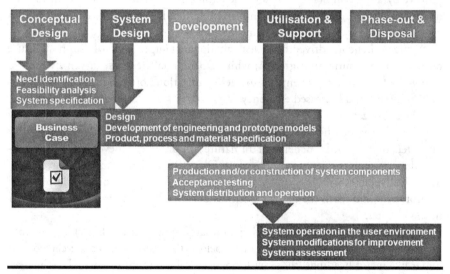

Figure 3.4   The Product Life Cycle.

Following the approval of the corresponding initiative and program/project, Business Benefits Management will proceed along the standard Product Life Cycle (Figure 3.4).

From the approved Business Case, the organization will launch the respective program/project solution to produce the Deliverables that then enable the initial operation to ramp up to full exploitation. In this period of utilization of the delivered solution, the operational unit will monitor and control its performance progress against the specified benefits.

Failures to realize Business Benefits are caused by flawed studies on feasibility, financial calculations, organizational impact, and an incomplete Business Case accompanied by invalid assumptions. Business Goals may be redirected, while uncontrolled evolution of Business Case goals, external market/business changes, and disruptions will all contribute to failed Benefits Realization.

Additionally, the organization might be ill-prepared or unprepared with unaligned or incompatible old-to-new processes and may lack a comprehensive plan for re-skilling or retooling, leading to resistance to change due to an ill-informed organization.

Moreover, programs/projects might be badly managed.

3.3.2 Description of Business Benefits

A business benefit is the outcome of an action or decision for change that contributes toward achieving one or more business Objectives. The benefit is considered positive by any or all Stakeholders participating in the change. The benefit may be quantitative or qualitative, and the improvement resulting from the change needs to be accompanied by a measurement mechanism.

Strategic Benefits, driven by transformational change, primarily support future business opportunities and growth, while Operational Benefits, driven by tactical change, will deliver improvements to the organization's operation through process improvements and increased efficiency.

The organization may also embark on speculative change initiatives, such as new avenues of growth, where the achievement of the benefits is considered a high risk but offers high rewards. Benefits of all nature can only be achieved once the corresponding program/project accomplishes its Objectives and produces its Deliverables. Those Deliverables will then enable the operational units to perform in the changed environment and reach the desired Business Benefits. Enablers are discussed further in this chapter.

Business Benefits are varied and span a large spectrum of characteristics from quantitative to qualitative. As explained below, these benefits can be categorized under five major headings: those with financial values, as for calculated, estimated, and rational financial values, and those which are qualitative and Intangible Benefits.

Of importance is the anticipated timeframe required for the business benefit to be realized. Benefits targeted for the short term, usually for tactical projects, would be less affected by internal or socio-economic and geopolitical changes, even though these may impact the goals to be achieved. Special attention is to be given to Business Benefits to be realized in both the medium and the long terms, as these will be transformational, and it is expected that the programs/projects of change will be performed for a period of more than a year, which may be extended to multiple years, and thus more affected by changes outside of the control of the organization.

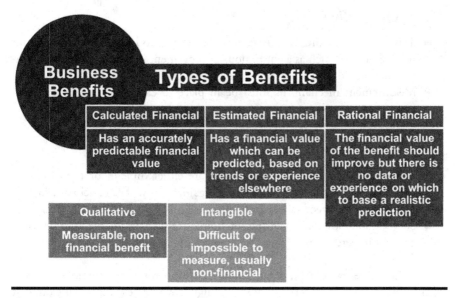

Figure 3.5   Types of Business Benefits.

Whatever is expected as the timeframe for the delivery of the results to be achieved for a business benefit, it is to be recognized that those benefits that are financially based are still speculative and thus approximate. It is also imperative that the funding required for the subsequent programs/projects to be launched is included in any financial analysis for the given type of benefit (Figure 3.5).

3.3.3 Types of Benefits

3.3.3.1 Calculated Financial Value

- It has a calculated and predictable value – such as the discarding of outdated equipment and the costs associated with its maintenance

3.3.3.2 Estimated Financial Value

- The value is estimated based on current expenditures and industry benchmarking. Confidence in this value is lower for this benefit, as assumptions made could be affected by internal and external changes

3.3.3.3 Rational Financial Value

- The value is an estimate derived following a rational and realistic prediction. As collected information and data or previous experience may be scarce or lacking, this financial benefit requires to be measured to monitor the improvement

3.3.3.4 Qualitative

- This refers to organizational changes that expect to bring an indirect nonfinancial benefit, such as a continuous improvement initiative, where the organization would gain on its effectiveness and efficiency
- Measurement mechanisms must be in place to capture operational performance enhancements

3.3.3.5 Intangible

- These benefits belong to changes that are either difficult or impossible to measure. These cover the changes the organization wishes to achieve, such as increasing employees' satisfaction, improving the image of the enterprise, etc.
- Surveys and other mechanisms of opinion are to be in place to measure changes in perception within the target group

Within the above framework, organizations must institute a measurement mechanism for all benefits, considering that the intangible would be exceptionally difficult to measure. Where target values for the calculated, estimated, and rational financial benefits are set, these must include calculations of Net Present Values (NPV).

Different types of initiatives would be launched to achieve Business Benefits, and programs and projects to be performed may need to be grouped for coherence and integrated into the corresponding Project Portfolio Management system. Groupings may be as follows, but not limited to:

- Profit margin enhancements
- Sales and revenue enhancement or acceleration
- Operational cost reductions
- Increase and higher productivity, including process improvement
- Improved customer quality of service
- Employee retention and satisfaction aligned with HR management
- Organizational policy or mandatory legal requirement
- Enterprise risk reduction
- Company image improvement

3.3.4 Benefits Management – Phases

Benefits Management is a structured approach. Too many organizations focus just on measuring improvements in the operational environment by deploying monitoring and control systems that capture the operational performance.

Organizations must consider that benefits can only be realized following a rigorous phased approach which is initiated at the business analysis stage and ends in the new operational environment.

Five principal phases constitute the structured approach.

3.3.4.1 Benefits Identification Phase

This phase is to be performed during the business analysis stage, when the organizational unit will establish the set of initiatives it wishes to pursue. The major focus will be:

■ Identify Business Benefits – what results the organization aims to achieve
■ Quantify Business Benefits – determine the type of benefits, set values for those that are financial, and describe those that are nonfinancial

3.3.4.2 Benefits Analysis Phase

This phase will select which initiatives are to be pursued and the programs and projects to be funded and launched.

■ Determine initiatives for each type of benefit, and set priorities
■ Expand the initiative to establish the set of programs/projects required to produce the benefits
■ Develop corresponding metrics measurement processes

3.3.4.3 Benefits Planning Phase

This phase consists of developing a holistic plan including both the program/project plan and its Deliverables, and the operational transition plan and performance following the delivery of the benefit enablers.

■ Establish a Benefits Realization Plan for both the program/project and the receiving operational unit. This plan includes total funding and resources, highlighting the major milestones to be reached
■ Institute benefits metrics measurement and monitoring processes
■ Construct the Benefits Map and enablers integrated into a master plan, which includes the program/project plan

3.3.4.4 Benefits Realization Phase

This phase consists of tracking, monitoring, and reporting the progress of the master plan.

■ Perform benefits plan monitoring
■ Record progress against set major milestones
■ Report status and forecasts for benefits

3.3.4.5 Benefits Transition Phase

This phase measures the Operational Effectiveness resulting from the delivery of the benefit enablers.

- Transfer Benefits Realization responsibility to the operational unit
- Capture operational performance on the delivered benefits

3.4 The Importance of the Business Case

As explained in Chapter 1, the Business Case is a mandatory step prior to any project launch. The more the business need is pertinent to sustainability and growth, the more the Business Case is to be comprehensive. Business Cases can range from a one-pager to a full and all-inclusive documented analysis of current and future situations that meet business needs (Figure 3.6).

3.4.1 The Business Case in the Overall Benefits Realization Process

The Business Case is an essential required step before the launch of a program/project. It responds to a transformational or operational need and justifies undertaking an organizational initiative.

❑ Preface
❑ Table of Contents
❑ Executive Summary
    ❑ Business Drivers – Goals to Achieve
    ❑ Recommendation
    ❑ Summary of Results
    ❑ Decision to be Taken
❑ Introduction
❑ Scope of Change
    ❑ Current Situational Analysis
    ❑ Internal & External Range of Scope
    ❑ Organisational Spread
❑ Financial Metrics
    ❑ Financial/Non-Financial Benefits
    ❑ Life Cycle Costs
    ❑ Funding Flow (NPV)
❑ Analysis
    ❑ Organisational Impact
    ❑ Assumptions
    ❑ Risk Plans
❑ Conclusion, Recommendation, and Next Steps
❑ Appendix

Figure 3.6 **Generic structure of a Business Case.**

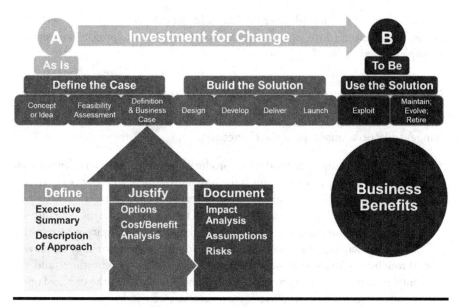

Figure 3.7   The Business Case documentation context.

The Business Case assesses the current and future organizational structures and proposes the scope of change required and establishes the financials for the ownership and Operational Benefits and costs.

The Business Case usually consists of two major facets: the Current Situation, which states the need, issue, and/or problem, and the solution proposal end situation, which presents the desired business outcomes and benefits.

A road map for the investment for change is established, following a classical Product Life Cycle scheme, to transition from the "As-Is" to the "To-Be", producing the program/project Deliverables and enablers, and reaching the operational environment where the Benefits Realization can be measured (Figure 3.7).

3.4.2 Financial Values as Drivers of a Business Case

Financial forecasts are estimates of future financial outcomes for the organization. A Business Case is used to evaluate the proposed initiative projects and decide whether to accept or reject them. It is a formal statement of a need to fund a given project and must be supported by solid business reasons. The Business Case defines the outcome, which is the result of the change derived from implementing the project, whereas the measurable benefit is achieved from the outcome.

The total funding to be made available must cover the total project, be it from CAPEX or OPEX sources, and includes all costs for project performance as well as expenditures related to the procurement of infrastructure components needed for the execution of the project.

Total funding for the initiative also includes all calculated and expected costs that will be engaged in the operational environment once the project Deliverables are implemented.

3.4.3 Calculating Cost of Ownership and Operations

As illustrated in Figure 3.8, funding is divided between ownership (create the change) and operations (exploit the change results).

- Ownership will require funding to produce the project Deliverables through to handover to operations – this is the initial investment to produce the results and make available the enablers to the operational environment.
- Operations funding consists of two parts. The first part concerns the ramp-up costs in implementing the results and enabling the organization to perform in the new environment and includes organizational alignment of employees through training, infrastructure replacements and/or modifications, and the implementation of new processes. The second part concerns the standard operational costs, such as employees, materials, services, and so on. Furthermore, operations funding must consider the timeframe for "parallel" performance of old and new operating environments, where the organization might have to function under two environments until a cut-off is satisfactorily reached, thus additionally training its limited resources.

Benefits can only be realized once the enablers have been implemented, and that occurs after the completion of the project Deliverables. It is therefore important

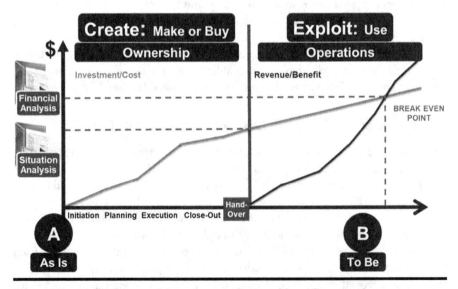

Figure 3.8   Cost of ownership and operations and benefits growth.

that the Business Case clearly presents the cumulative cost curves covering both the ownership and the operations periods, and the estimated cumulative benefits, which can only be measured from "day 1" of operations.

For the financial value benefits, the Business Case will illustrate key indicators such as the break-even point, Payback Period, and ROI. These must be calculated with Net Present Value.

For qualitative and Intangible Benefits, the benefit cumulative curve cannot be other than speculative.

3.4.4 Major Steps to Validate a Business Case

Key business drivers as identified in the different Business Cases must be evaluated, and the corresponding benefits must be valued. Benefits to be realized must be substantiated for the Business Cases to be accepted and to proceed to ranking and setting priorities.

Financial projections are of vital importance and must be finely calculated and illustrated to render acceptable and coherent the Business Case and the subsequent initiative that will be launched. The projections must quantify the benefits logic and seek validation with key Stakeholders, and as a minimum cover the following:

- Financial baseline – overall ownership and operations costs/revenues
- Financial and operational assumptions
- Break-even, ROI, and Payback Period assumptions

The valuations will establish a ranking according to the organization's priorities, differentiating into two groups the transformational and the tactical Business Cases. The former may well have a higher priority in the organization's overall Objectives, and thus relegate tactical Business Cases to a lower position in their ranking.

Within the grouping of transformational and tactical Business Cases, subgroups of tangible and Intangible Benefits can be established to refine ranking and priorities.

Tangible benefit realization goals can be grouped under "sustainability" or "growth", while the Intangible Benefits can be classified under "organizational". The subgrouping can be further structured as shown below, but not limited to:

- Company image
- Cost reduction
- Customer service enhancements
- Environmental sustainability
- HR – employee retention
- IT infrastructure improvements
- Legal or statutory/regulatory requirement (mandatory)
- New product/service introduction
- Productivity increase

- Process improvement
- Revenue enhancement or acceleration
- Risk reduction
- Supply chain management
- System rationalization
- Total quality management

Once the Business Benefits described in the Business Case are confirmed, ownership of the benefits is assigned to a corresponding organizational unit that will be pivotal to their realization. A Benefits Map is developed, as is explained further in this chapter, and is integral to the program/project that is launched.

The operational unit will develop and institute the Organizational Readiness plan to implement the initiative and then operate and deploy an agreed-upon Benefits Realization tracking process that is assessed against the Business Case goals and Objectives.

3.5 Programs as Agents of Benefits Enablers

Please refer to Chapter 2, "Management of Programs", and Chapter 5 for details on Project Portfolio Management and the PMO for a detailed discourse on these two topics.

In Business Benefits Management, organizational business KPIs can only be achieved and measured during operational performance post each project close-out and handover. This is established principally by:

- Clear definitions of KPIs as business strategic value drivers
- Ensuring KPIs are established on validated information and are coherent and understandable
- Distinguishing between financial and nonfinancial KPIs
- Determining the "depth" of the cascade of the KPIs to other areas of the organization

The business KPI covers the total Product Life Cycle, whereas the program KPI is contained within the boundary of its solution-building development cycle (Figure 3.9).

To ensure minimum confusion and maximum misunderstanding by all parties, the program and Project Managers must be intimately knowledgeable about the organization's business KPIs.

3.5.1 Consolidated Program Benefits

Because of the different initiatives that have been and are to be launched, the organization must be rigorous in the ranking and priority-setting method it employs (Figure 3.10).

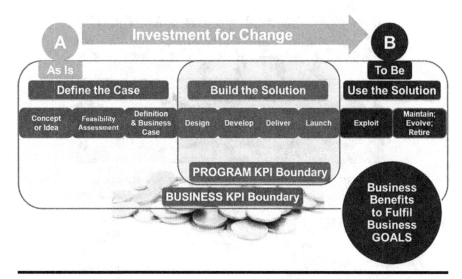

Figure 3.9   KPI boundaries.

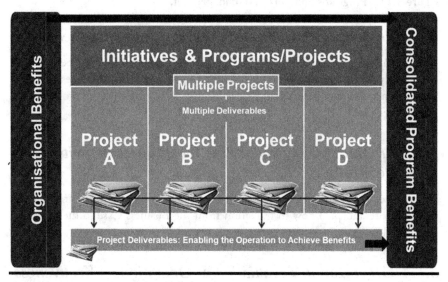

Figure 3.10   From initiatives to programs and projects.

The organization must consider how to mix both transformational and tactical change initiatives. Programs and projects that are performed without sufficient assessment of funding and resource requirements across all other initiatives will cause conflicts that will prohibit the organization from achieving any, or all, its Objectives and Benefits Realization.

Figure 3.11   Programs and portfolio management.

3.5.2 Program/Project Strategy and Portfolio Management

It is also imperative that the Project Portfolio Management system be deployed accordingly and that the organization have a robust mechanism to assess the total funding and resources it needs to achieve its goals of change through programs and projects (Figure 3.11).

3.5.3 Program Management Plan for Business Benefits Realization

When selected, the chosen program/project launched to achieve the desired benefits will structure and plan how to transition from the Business Case to the operation (Figure 3.12).

3.5.4 Organization's Evaluation of Operational Effectiveness

The Program Manager will define, in collaboration with the respective business units, how the change initiative will be delivered and how and what the operational unit must institute and deploy to enable the realization of Business Benefits in its area.

To facilitate how the operation will be capable of evaluating its new Operational Effectiveness, a structured plan is developed that covers the business Objectives, Stakeholders, changes in the operation, and the required change enablers.

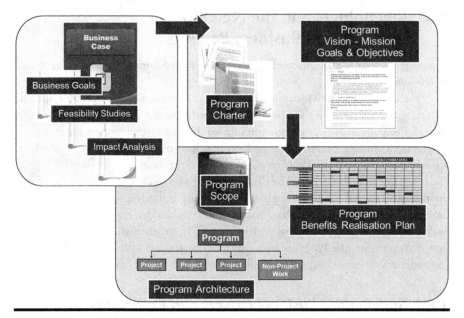

Figure 3.12   Synopsis of a program plan for Benefits Realization.

3.5.4.1 Business Objectives

■ Defining high-level priorities in relation to the drivers, outcomes, and improvements to be delivered upon the completion of the project

3.5.4.2 Business Stakeholders

■ Identification of all individuals or groups who will benefit from the project and are either affected by or directly involved in making the changes needed to realize the benefits

3.5.4.3 Business Changes

■ Description of all business activities and new ways of working that are required to ensure that the desired benefits are realized

3.5.4.4 Enabling Changes

■ Definition of the enablers as prerequisites for achieving the business changes that are essential to bringing the new system into effective operation

Please refer to Chapter 2 and Chapter 5 for details on Project Portfolio Management and the PMO.

3.6 Stakeholders and the Need for the Realization of Business Benefits

Please refer to Chapter 4, "Stakeholder Management and Engagement" for more details on this topic.

Stakeholders are individuals or organizational entities who represent specific interest groups served by the outcomes and performance of a program/project of change. Program Managers are accountable for the end-to-end management of their programs, including performance and expectation management of individuals or organizational entities who may be outside their direct control.

The major Stakeholders will be the business drivers of change and seek to achieve their organizational goals through the launch of initiatives and programs. The Program Manager must learn and incorporate stakeholder expectations and fully understand which organizational entities benefit from successful program completion. Furthermore, the Program Manager must determine which Stakeholders have contradictory interests that could conflict with the program's interests.

3.6.1 Stakeholder Business Drivers and Expectations for Strategic Program

As detailed in Chapter 4, Stakeholders have different concerns at different stages of the program life cycle (Figure 3.13).

Business drivers refer to those Stakeholders who are the driving force behind the program of change, have a strategic view of the required change, and are set the business benefit expectations for the delivery of value in operational use.

Project performers refer to all individuals and entities, internal or external, who will perform the program and project activities. They will have a tactical view focusing on project/task delivery and expectations to produce project Deliverables.

Operational performers refer to operational entities and employees who will enable the delivery of Operational Benefits. They will have an operational view, and expectations are on usable and exploitable program/project Deliverables.

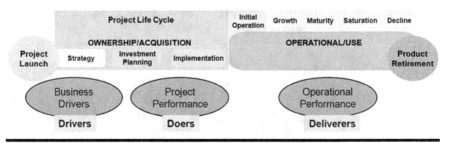

Figure 3.13   Stakeholders and the program life cycle.

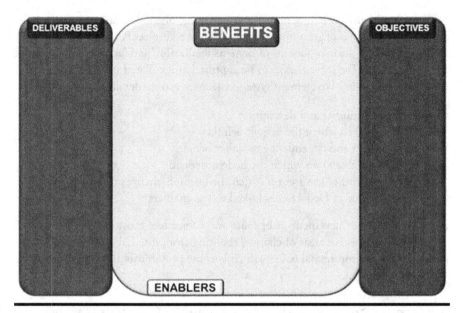

Figure 3.14   Benefits Map template.

3.7 The Benefits Map

The Benefits Map is an extremely valuable tool as it illustrates the Objectives to be achieved, the links to realizing the Objectives of the benefits, the positioning of the enablers to the benefits, and the program/project Deliverables that drive the map (Figure 3.14).

The flow describes a series of cause-and-effect relationships from project Deliverables to enablers, and consequently to benefits and the Objectives to be achieved.

The initial Benefits Map is generated in the benefits planning phase (see Section 3.3.4.3 above) and is provisional. As the map construction proceeds, Deliverables and enablers are linked and connected to benefits and Objectives, thus representing a map with a set of viable associations.

3.7.1 Process of Identifying and Structuring Benefits

The identified Objectives to be achieved relate directly to the organization's declared Vision and Mission and the consequently launched initiatives and programs/projects. Each objective will generate a set of benefits, which will probably be related to one another. Certain benefits may be achieved earlier than others, creating connections between them to lead to the Objectives.

Objectives are decomposed into end benefits and lead to subsequent decompositions of these benefits into contributing benefits, which are related to the enablers that flow into them.

Effective identification, analysis, and grouping of a large quantity of benefit-related information is of great importance. A deployed Project Portfolio Management system collates all the components of Benefits Realization and facilitates analysis and communication. The information to be captured and collated in conjunction with the Project Portfolio Management system will cover various details about the benefit:

- Reference number and description
- Objective(s) to which the benefit is linked
- Relationship and dependency on other benefits
- The Deliverable(s) on which the benefit depends
- Enablers or other features on which the business changes depend
- Program/project Deliverables linked to the enablers

Risks for the nonachievement of benefits and Objectives must be assessed, as well as concerns in sensitive areas of change, resulting from internal and external social, political, and environmental issues, which become prohibitors and inhibitors to the whole Benefits Map.

3.7.2 Benefits Realization Plan and the Benefits Map

The Benefits Map represents how program/project Deliverables relate to the Objectives to achieve and is updated responding to changing business needs.

However similar, Benefits Maps and process diagrams are not the same. Benefits Maps describe the cause and effect between its enabler components and the business changes needed for the delivery of the benefits. Project or program planning will subsequently incorporate the tangible Deliverables to be produced for the enablers.

The Benefits Realization Plan is developed in conjunction with the Benefits Map, and consists of:

- Identification of the expected benefits
- Ownership of the Benefits Realization
- Measurement methodologies for Benefits Realization
- Benefits delivery schedule and interfaces to operations
- Procedures for ongoing Benefits Realization in operations

If the Benefits Realization Plan and the associated Benefits Map are created prior to the availability of detailed program/project plans, the plan and map are to be reviewed for potential conflicts in the flow of Deliverables to enablers and benefits.

3.7.3 Developing the Benefits Map

A Benefits Map is initiated by positioning the Objectives on the right-hand side. End benefits that directly relate to the Objectives are then inserted and linked to their corresponding Objectives. End benefits are reviewed with all the concerned Stakeholders to reach an agreement that the objective(s) can be fulfilled once the end

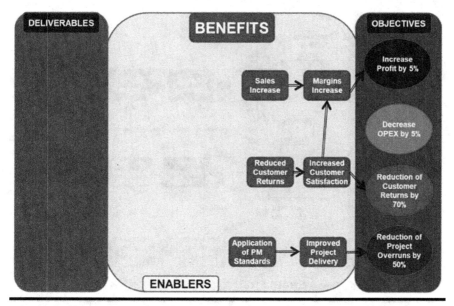

Figure 3.15   Positioning Objectives and benefits in the Benefits Map.

benefits have been achieved. Further testing can be performed to determine if the end benefit is correctly positioned, by removing it and assessing if the objective would still be fulfilled. If the removed benefit is not an end benefit it would likely be an inter-mediate/feeder benefit, which will be linked to another end benefit (Figure 3.15).

The Benefits Map construct proceeds by identifying intermediate/feeder benefits and creating the respective links, either directly to the end benefit or to each other.

At this stage and once the Benefits Map has been populated with Objectives and benefits (end and intermediate/feeder) and an agreement has been reached, the enablers and the business changes needed for delivery of the benefits are identified and placed to link to the corresponding benefits (Figure 3.16).

When additional changes may be necessary to ensure coherence in the flow lead-ing to benefits, additional enablers are inserted in the Benefits Map. For example, it may be that an "employee re-training" change on a new system is necessary to lead to the change for "productivity improvement".

Program/project Deliverables are then inserted in the map and the links to the enablers are established, which concludes the map (Figure 3.17).

The completed map is reviewed by all the concerned Stakeholders, modifications and adaptations are made, and an agreement is reached.

■ Example: OBJ03, "Production and Manufacturing: Rationalization Program", consisting of four projects (Figure 3.18).

The author recognizes the work of Gerald Bradley in Benefits Realization Management.

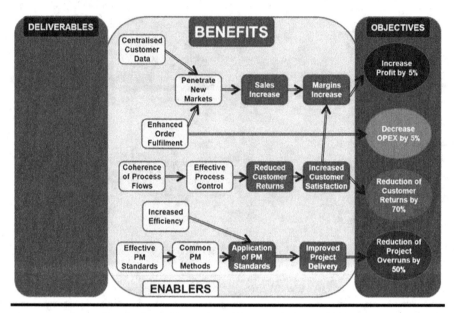

Figure 3.16   Positioning Operational Enablers in the Benefits Map.

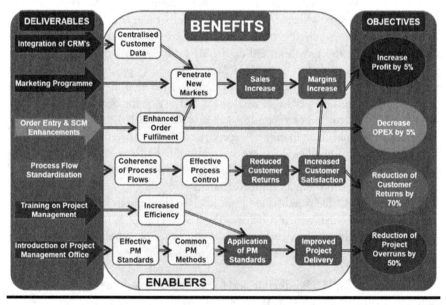

Figure 3.17   Benefits Map completed with Deliverables associated with the enablers.

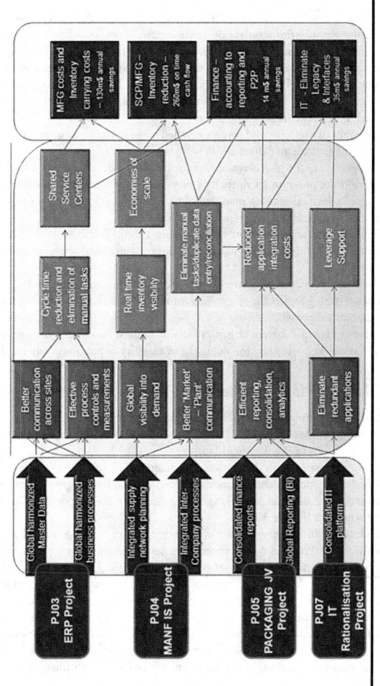

Figure 3.18 Example: OBJ03 – Production and Manufacturing: Rationalization Program.

3.8 Analysis of Program Delivery of Benefits Enablers

The Benefits Map construct depicts the sequence of changes that will be incorporated into the organization's operating environment. Anticipated projected dates placed in the map as to when enablers and benefits are targeted to happen depend solely on the program master plan, as those changes can only intervene once the program and project Deliverables have been produced. It is therefore essential for the organization to focus in detail on the progress of work during program execution as described in the benefits planning phase.

The program master plan and project detailed plans must demonstrate the value of the investments made to deliver the benefits.

The program master plan development will include details from each of the projects within the program (at times that may be just the one project) and consist of the project's detailed activity plans, resource needs, scheduling with milestone highlights, budget spending, and resulting Deliverables.

Projects in execution will report on the progress through the established monitoring/control mechanism and signal their project closures and the availability of their Deliverables.

The program master plan aggregates the individual project plans, and updates are conducted during and at the end of each project, where original milestone dates for Deliverables are revised.

The master plan also covers Organizational Readiness and operational start-up activities that can be performed following the availability of program Deliverables.

3.8.1 Benefits Planning Phase for Benefits Realization

Key milestones and project Deliverables are aggregated into the program master plan and relate the schedule of the Deliverables to the business Organizational Readiness preparation and thus to the Benefits Map.

The program master plan drives the performance of the change execution. The master schedule is updated and consolidated according to the progress of the projects it covers, and evolutions and changes are reflected in the program and project scope and schedules (Figure 3.19).

When internal and/or external factors arise that impact the program and project scope of work, the Program Manager team reassesses the Business Case, conducts an analysis of the work to be performed, and reviews the Deliverables schedule. The program master plan is accordingly adjusted, and the results are reflected in an updated benefits/Deliverables target date schedule in the Benefits Map (Figure 3.20).

The operational unit concerned with the change program and the project's Deliverables prepares a schedule for the Organizational Readiness activities in its area of responsibility. The schedule that will initiate the enablers in the Benefits Realization Plan in operations is developed in collaboration with the project team (Figure 3.21).

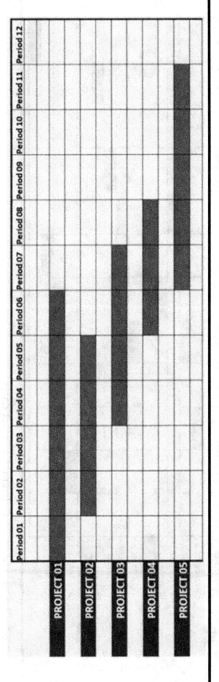

Figure 3.19 Program Master Schedule.

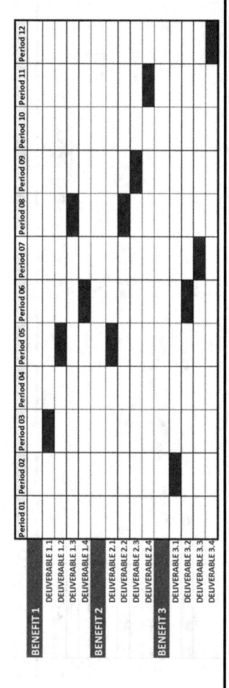

Figure 3.20 Program benefits/Deliverables target dates.

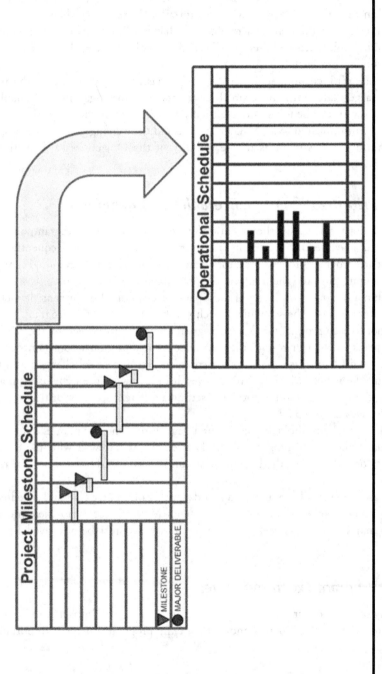

Figure 3.21 Project Deliverables milestone schedule interface to operations.

As project Deliverables are transitioned to the operational unit, the mix of projects within the program may evolve as organizational needs may change following the deployment of the delivered results. This has implications for the remaining Deliverables' milestones and their previously dated links to the enablers for Benefits Realization.

All changes incorporated in the program master plan will impact the provisional target dates initially set in the Benefits Map for triggering the enablers and the flow to the benefits and eventually terminating at the Objectives. This must be communicated to all Stakeholders, as radically changed target dates may well require an assessment as to the viability of the Objectives to be achieved (Figure 3.22).

3.8.2 Prioritization of Deliverables and Benefits

Originally set priorities for the execution of projects within the program, and the anticipated schedule for the creation of deliveries, are to be reviewed frequently during standard program progress monitoring and control to ascertain required changes to the program plan and the resulting Benefits Map.

Within a program consisting of multiple projects, not all projects are launched on the same day, and the schedule of each will be reflected in the program master plan. Prior to each project launch, the Program Manager team reassesses the original project plans to align or realign these to cater to organizational constraints, project scope and Objectives evolutions, schedule and resource need changes, and the impact on the master schedule. This will result in reviewing the Benefits Map and applying the necessary modification and setting a different set of priorities to reach the Objectives (Figure 3.23).

Additionally, Stakeholders may request that during the program execution of individual projects, the progress of the Deliverables is reviewed with the benefit owners to determine how the Benefits Map can be rearranged to allow for the realization of "quick-wins".

The assessment will produce an up-to-date program master plan and schedule.

All changes to be applied to the Benefits Map to reset priorities, including the map's element contents, structure, and flow must be approved by the Stakeholders.

3.8.3 Program Execution Issues

A variety of issues occur during program execution and their impact is to be measured not only on the progress and status of the program but more importantly

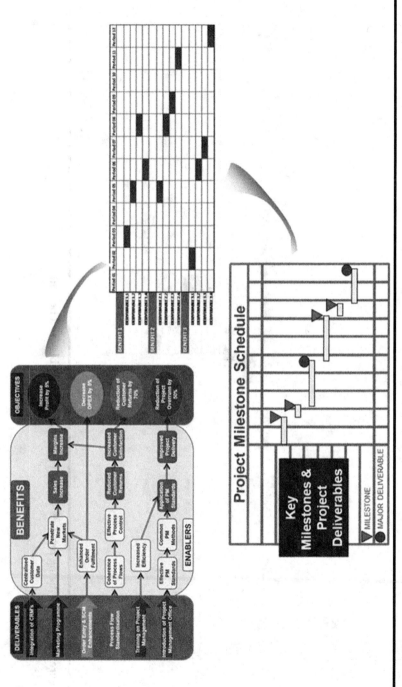

Figure 3.22 Synchronization of Deliverables, program schedule, and Benefits Map.

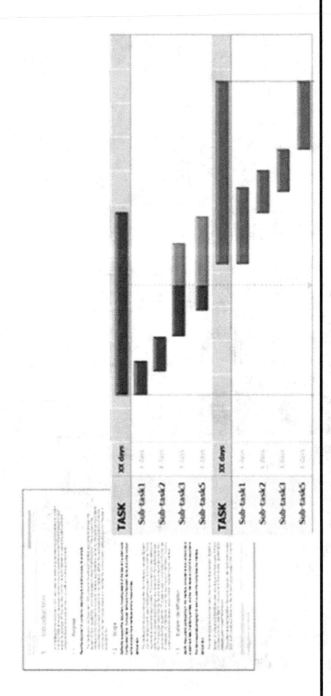

Figure 3.23 Prioritizing projects/tasks in line with Deliverables and benefits.

against the Benefits Realization Plan and map. A non-exhaustive list of common issues to anticipate is discussed below.

Stakeholder Management:

- Evolving scope – internal and external environments, be them organizational or market needs, will result in Stakeholders reassessing the original Business Case Objectives and making scope content changes as they deem required
- Change requests – from the evolution of scope and from other factors related to the program or project, such as obstacles encountered by a contractor, change requests will be generated that, if approved, may modify the program or projects significantly and the overall Benefits Realization scheme
- Business Organizational Readiness – this refers to issues encountered by the receiving operational unit, in that the necessary change adaptations to employee skills, tooling, and infrastructure and modified/new processes, performed by and under the responsibility of the unit, are delayed and/or are no longer relevant

Organizational and Operations:

- Business Case changes – in medium- to long-term duration programs, the Business Case will be constantly under scrutiny to assess consistency to its original intention
- Operational blockages – these would usually occur from a lack of Organizational Readiness preparedness, business unit stakeholder opposition, unit personnel resistance to change, trade union strikes, and operational launch timeframe conflicting with the end-of-fiscal-year closures
- Funding restriction – the operational unit is funded by its annual budget, OPEX. Maintaining and sustaining current operational performance will often have priority over programs/projects of change, as financing both may be limited within the fiscal year

Financial and Regulatory:

- Economic instabilities – be they internal, due to revenue and productivity loss, or external, due to evolving market conditions, resulting in funding restrictions for the planned Objectives
- Changing norms – affecting the organization's current products/services and regulatory changes to the way it operates, requiring a review and a reassessment of its processes and necessitating funding changes
- Natural disasters – all calamities affecting the organization and its environment

Technical:

- Design issues – with respect to or in connection with the engineering specifications

- Testing problems – incorrect or unstable test environment, a lack of skilled testers, incorrectly estimated testing schedule
- System component defect – deviations or irregularities from the specifications or statement of work encountered during testing, from either internally or externally produced components
- Integration difficulties – late arrival of system components, incomplete integration and commissioning plans, the lack of skills of integrators, and the inability to reproduce the target operational environment
- Quality of Deliverables – as a result of all the abovementioned points

Schedule:

- Incorrect estimates – resulting in slippages in delivery, additional resources, and supplementary funding
- Resource unavailability – inability to perform work according to the plan, creating a cascade effect of delays for successor activities
- Materials/equipment delays – due to market shortages, supplier schedule slippages, customs clearance issues
- Contractor performance and delays – nonquality of work performed, inaccurate delivery schedule estimate
- Users unavailability – operational unit employees unavailable during integration and commissioning tests and either absent or engaged in their unit's performance schedule and not receiving the appropriate training on the new changes

3.8.4 Focus Areas for Benefits Realization

The program/projects deliver a functional solution and support it. The operation exploits it to deliver the Business Benefits.

Benefits can only be realized in the operational environment, where enablers of change flow into intermediate benefits to terminate at the end benefits and corresponding objective(s). For the operational unit to be effective, the preparatory Organizational Readiness must have been completed and the employees, infrastructure, and processes must be available and engaged to perform in the new environment.

It is to be noted that delays in certain project Deliverables will hinder the initiation of an enabler and prolong the overall timeframe in the flow from the enabler to the final objective.

The operational unit in conjunction with the program management team will be deploying an agreed-upon objective measurement mechanism, instituted in the operations environment to monitor and evaluate progress toward Benefits Realization (Figure 3.24).

The measurement method must demonstrate that the organization is enabled to realize the Business Benefits. The measures recorded will be financial for the quantitative

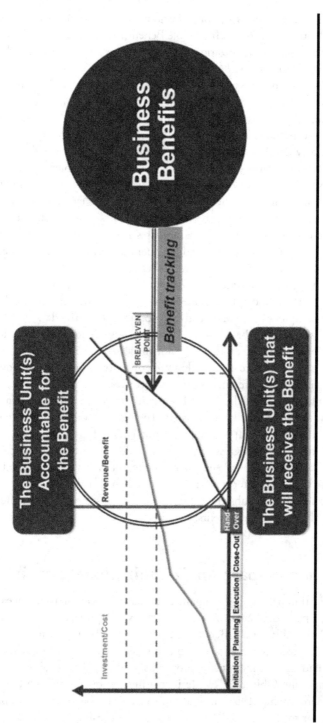

Figure 3.24 Measurement of Benefits Realization operations.

and tangible benefits, whereas the organization must decide on what measures are acceptable and satisfactory for the Intangible Benefits.

Of importance to the organization is to ensure that all costs associated with the programs/projects implemented are accrued, to which the operational costs will be added, thus allowing for assessment of the benefit value in its realization.

The operational unit progressively ramps up its operation by performing the enablers to lead to the delivery of the Business Benefits. There are target values to achieve within a timeframe in operations when improvements are expected to begin, and a timeframe when the break-even point is expected to be reached.

The initial focus is on the deployment and use of the enablers by the performing unit. These will be the processes and new ways of working, coordinated with the new employee skills and the planned infrastructure evolutions. The enabler scheduled dates, for when the change improvement is expected to start and end, are monitored and controlled. Performance measures are recorded at an appropriate frequency, from hourly to weekly or monthly depending on the nature of the enabler, to compare initial baseline targets to achieved targets.

As enabler progress is recorded, the Benefits Map is updated to reflect when the measurable benefit value can be initiated and consequently achieved and the estimated date of completion of the objective(s) accordingly rectified.

The organization must be ready to make decisions following the performance measurement of enablers when these do not meet expectations. This would entail deciding on which enabler-to-benefit flows to concentrate on, or those flows which are no longer viable, thus affecting the end Objectives.

Once enablers are deemed to have been completed, the values of benefits realized can begin to be measured and assessed. The planned values, dependent on the completion of the enabler that preceded, are not achieved instantaneously. The benefit value would (should) progressively increase following the performance of work, for example, a process improvement or a marketing sales drive. The tracked progress of the benefit's value will determine if the originally planned Objectives to meet are still feasible, and this may subsequently involve a reassessment of the Benefits Map and the derived values of the benefits.

3.9 Plan and Prepare for Organizational Readiness

Benefits Realization will fail if the organization has not been prepared to exploit the change in the operational environment. Programs and projects may have delivered exploitable Deliverables and a Benefits Map may have been produced to illustrate the flow from enablers to benefits and to end Objectives. However, if the organization has not laid the foundation for proper usage of the Deliverables, then it will struggle to meet the Objectives, while increasing the operational costs of current and future environments. This will result in investments, be they CAPEX or OPEX, that are superior to those initially envisaged, as the break-even point according to planned dates may not be reached, or even never attained, and ROI and other financial measures will also not be realized.

It is therefore imperative that the Organization Readiness be conducted as a project by the operational unit, and its plans integrated into the overall change plan of programs/projects.

3.9.1 Planning the Organizational Readiness

The organization must ensure active sponsorship at the senior executive level and engage this sponsorship for a successful readiness transition.

The Program Manager and the team should coordinate with the business unit(s) to discuss the details of the readiness plan, involving the right people in the design and implementation of changes and securing buy-in from those involved and affected.

The readiness plan must include actions to assist the people affected and impacted by the change, to be ready to adapt to the changes by ensuring they have a comprehensive understanding of how the plan will be beneficial for them.

A large portion of the readiness plan will be devoted to training the appropriate business unit resources on the changes to processes, procedures, methodologies, and systems. The readiness plan may also include work performance for the creation/ adaptation of all required new processes, or this may be a project within the Program Scope. Similarly, infrastructure work performance, such as materials and equipment additions or renovations, IT systems, and other operationally related systems, may be included in either the readiness plan or the program plan. The importance of the decision will determine accountability.

3.9.2 Preparing for Organizational Readiness

The program team and the business unit(s) work in harmony to communicate, inform, and explain to the functional departments to establish awareness and understanding of the changes required and gain acceptance and commitment.

Awareness of:

- Goals, purpose, and rationale
- Business Benefits and need for change
- System to be implemented
- Timeline and intended audience

Understanding the:

- System and business process changes
- Impacts of business process changes
- Key milestones
- Training planned to be provided

Acceptance of:

- Implementation schedule progress
- Issues and resolutions
- Training schedule
- Implementation and transition schedule

Commitment to:

- Roll-out and ramp-up activities
- Support processes and follow-up activities
- Business Benefits measurement mechanism

3.9.3 *Conducting the Readiness Assessment and Plan*

Organizational Readiness Assessment determines the extent of the multiple and simultaneous changes required to be made in the operational environment. The readiness plan, conducted by the program team in conjunction with the business units, establishes the activities to be performed and assigns responsibility and accountability. The readiness plan has three major focuses: people, processes, and infrastructure.

People:

- Organizational chart changes
- Staffing requirements
- Training for re-skilling
- Recognition and reward system

Processes;

- Definition and development of new/modified procedures and methodologies
- New or modified workflows
- Internal and external changes in communication paths
- Governance and decision-making during and after the readiness is completed

Infrastructure – Additions and/or changes to

- Property assets
- Production assets
- Technology platforms
- Transition and migration of legacy data

Important to note that the Organizational Readiness Plan will be performed in parallel with normal operations. It is therefore essential for the operational unit to cater to the required resources during this phase. Readiness may continue in conjunction with the start-up and ramp-up of the new operational environment and changes to processes and employee skills may hinder operations performance. Furthermore, operations must manage its fiscal year OPEX funding, which is divided between normal operations and the performance of readiness activities, especially when the transition is governed by a transformational change initiative.

3.9.4 Resistance to Change

Projects create change. An ill-prepared change transition will create resistance to the change. Obstacles will also arise because of incomplete or ineffectively managed readiness and transition to the operations. These obstacles will create inhibitors to enabling change and will lead to organizational and individual resistance. The major causes are itemized below, as a non-exhaustive list.

Organizational resistance will be met because of:

- Threat to established resource allocation and distribution
- Threat to established power relationships
- Threat to competencies and skills
- Scope and extent of change
- Structural lethargy

Individual resistance will exist when employees are confronted with:

- Fear of the unknown
- Loss and security
- Moving out of a comfort zone
- Economic factors

Overcoming resistance is not an easy task. Management must involve all operational employees and line managers/supervisors who will be affected by the change from the launch of the change initiative. Participation of these individuals in the design and planning of the change will greatly enhance the coherence and the plan to transition to the new operational environment. The program master plan must schedule workshops with all organizational entities that will use, exploit, and/or operate in the new/changed environment, and adjust its contents accordingly following feedback and input from operational employees.

3.10 Program/Project Transition and Handover

The program team and the corresponding operational unit(s) will conduct a transition and go-live assessment and seek confirmation from key Stakeholders for the policy and governance to use for the achievement of benefits and goals. The assessment will cover:

- Review of workflow and process improvements
- Review of staff schedules
- Review of all user documentation
- Training delivery
- Collation of all vendor support contracts
- Method for the evaluation of results and evidence of testing

3.10.1 Transition and Go-Live Challenges

The major challenge with which the organization is confronted is the interface between program delivery, Business Change Management, and operations.

The transition follows a plan and refers to the progressive migration and switch from the old to the new operational environment, while go-live refers to the first day the new system will be used in a production mode by at least one user.

The transition plan may be a subset of the Organizational Readiness Plan or be a specific operational stand-alone plan. When required, the stand-alone plan is established when the operations transit from old to new environments over a lengthy multi-month timeframe which will include the parallel performance of old and new systems. Enablers of change will occur during the transition, and flow to the benefits to be realized; however, the operations will be required to effectively plan its surcharge of resource needs.

The transition will require a realistic schedule that takes into consideration the workload requirements of the operational unit as it pursues its regular performance of maintaining business operations. The transition plan will interface with the readiness schedule and prepare the business and functional departments for the go-live day.

Go-live will be initiated once the operational unit has:

- Organized appropriate support team staffing
- Ensured provision of help desk staff, super users, etc.
- Provided disaster recovery business resumption plan
- Instituted escalation and approval procedures
- Coordinated the schedule with vendors, contractors, and other providers

3.10.2 Enablers of Change and Realization of Business Benefits

The performance of enablers in operations is tracked and monitored from day 1 of the go-live day. The new mode of operation in the environment implements and utilizes the established objective measurement method and validates the data across all investments to demonstrate that the results enable the delivery of business value.

Performance is measured against pre-established criteria and decisions are made when and where deviations exist. Upon the completion of an enabler or a set of enablers, measurement focuses on the benefits accrued and eventually realized. As has been stated previously in this chapter, benefits are not instantaneously realized.

3.11 Conclusion – Business Benefits Realization

Organizations strive for value, and the purpose of change should always be the realization of Business Benefits. Organizational Objectives drive initiatives that are converted to programs and projects that produce results and Deliverables. These enable operations to function in a new or modified way and deploy resources to produce the necessary changes to attain the Business Benefits and achieve the organizational Objectives and goals.

In conjunction with program and project plans, Organizational Readiness, and operation transition plans, the organization establishes a comprehensive set of Business Benefits to achieve and structure a Business Benefits Map.

The litmus test for the complete scheme is the effectiveness of the organization to engage in a comprehensive and coherent approach to Business Benefits Realization.

Chapter 4

Stakeholder Management and Engagement

Please also refer to Chapter 2, "Management of Programs", for further discussion on the role of Stakeholder Management in the Management of Change

4.1 Chapter Overview

Organizational Change Management will always involve Stakeholders, at the executive level for transformational changes seeking growth and opportunities and at the tactical functional level for changes to sustain operations and maintain a competitive advantage.

Stakeholder Management is an important skill for all Project Managers. This chapter describes the Stakeholder Management tools and techniques most pertinent to deploy throughout a project, and a large part is dedicated to the specific focus to be placed on the personal and communication skills to be employed by the Project Manager.

Changes create programs and projects which will have impacts on:

- Organization structures
- Policies and procedures
- Job responsibilities
- Communications
- People – human resources
- Motivation of employees
- Cultural evolutions
- Recognition and reward systems

DOI: 10.1201/9781003424567-5

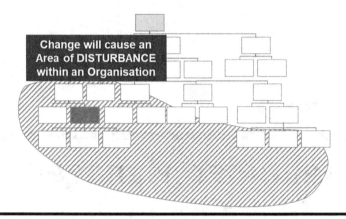

Figure 4.1   Scope of change disturbance in the organization.

The size and scope of any change will affect parts or the whole of the organization. This will create potential resistance to change, which is a natural human condition, as people move out of their comfort zone and seek to overcome their fear of the future (Figure 4.1).

Raising organizational comfort levels is a primary consideration in the Management of Change, making the future real, tangible, and deeply understood in people's minds. At certain times, change is absolutely necessary as the Current Situation is deemed to be completely unacceptable. Conducting the change process is to engage in a participative and rigorous manner that takes many perspectives, involves many people, and builds a solid understanding of why the change is needed.

All the people mentioned above are Stakeholders, and for any project of change to be successful, Stakeholder Management is the pivotal key success factor.

Stakeholders are most commonly defined as individuals and organizations, be they internal or external to the enterprise.

PMI defines Stakeholders as Individuals and organizations who are directly or indirectly involved in the project, or whose interests may be positively or negatively affected as a result of project execution or successful program/project completion.

Stakeholders are individuals who represent specific interest groups served by the outcomes and performance of a project or program. Project Managers are accountable for the end-to-end management of their projects, including performance and expectation management of individuals who may be outside their direct control.

The Project Manager is not the only person to address the organizational harmonization for change, as senior executives and the project sponsor have that overall responsibility. However, as the project is the central visible change enabler, the Project

Manager is placed in a crucial role to meet the organization's stakeholder concerns beyond the focus of delivering the project to its scope contents and Deliverables.

Establishing the full scope of work and building stakeholder commitment from all parties will further enhance adherence to the change, while ensuring that stakeholder issues and interests are integrated into the scope of the change.

4.2 The Role of Stakeholders

Stakeholders are directly involved when the core scope contents are driven by their needs and/or when they will be recipients of the project outcomes. Stakeholders will be directly involved when resources assigned to the project are provided as per their responsibility areas. Those Stakeholders that are indirectly involved in the project will be peripheral in either the contents of the project scope or the deployment of the project's results. However, these somewhat distant Stakeholders are to be included in the total scope of the change as their operational participation in exploiting the project's Deliverables is a critical success factor. As will be discussed below, importance must be given to stakeholder interests as these may positively or negatively affect project execution and successful project completion.

Project Managers must give due consideration to the people issues surrounding projects and recognize that the appropriate involvement and management of Stakeholders is a critical success factor. Project Managers should therefore institute a formal Stakeholder Management process that is appropriate for the nature of the project.

The degree to which a stakeholder is required to contribute to a project to ensure success depends on a number of factors, including the stakeholder's position and authority within the organization, the degree to which the project is reliant upon the particular stakeholder to provide a product or service, the level of "social" influence of the individual and the degree to which the individual is familiar with specific aspects of the business. Each of these factors can be analyzed in more detail during Stakeholder Analysis.

4.2.1 Stakeholders' Expectations and Project Scope Management

Stakeholders have expectations to be met and needs to be satisfied by a project. Simply said, establishing and managing stakeholder expectations is to deliver the "right" project to the "right" people for the "right" business need.

However, not all Stakeholders are aligned with the same Objectives and goals.

Executives and line managers have positive/negative financial or emotional interest in the outcome of their performance for the organization and their interests will often go beyond that of the project. Additionally, their professional and personal interests will influence their decision-making and may create inconsistency in maintaining the boundaries of the project scope.

Line managers of functional departments may face conflicts when deciding to place a priority between meeting their operational goals and maintaining focus on the project scope.

External providers, suppliers, contractors, and vendors will place a great deal of scrutiny on the contours and contents of their respective statements of work and seek to safeguard their own organization's interests.

Government agencies and associated bodies will impose strict adherence to needs that the project must include in its scope.

Project scope management consists of the processes required to ensure that the project includes all the work required, and only the work required, to complete the project successfully. This is therefore not only complicated to establish but also complex, as Stakeholders, as mentioned above, may well disrupt the project scope according to the decisions (unilateral at times) they make on expectations and needs.

Project scope management comprises scope planning, scope definition, and scope change control, terminating with scope verification, which formalizes the final acceptance of the project scope.

At the start of a project, the enabling Business Case is reassessed to establish that the current expectations and needs remain consistent with those initially registered. As the Business Case may not have established the complete set of project Stakeholders, a more detailed analysis is to be performed during project initiation. The identified Stakeholders will be invited to describe in detail both their expectations and needs, through interviews, questionnaires, and other fact-collection means.

Stakeholder needs drive the initial contents of the project scope. An assessment of the captured needs is conducted to determine if conflicts and irrelevance exist and which of these needs is to be resolved or filtered out. The associated needs and expectations excluded from the initial project scope are communicated to the respective Stakeholders, thus alleviating any subsequent disappointment from these individuals.

The agreed-upon scope contents document, gathered from approved expectations and needs, is the principal input to the project front-end engineering and design stage.

4.3 The Need for Stakeholder Relationship Management

Project Managers navigate in a complex environment and spend between 75 percent and 90 percent of their time communicating orally and verbally. The communication paths in a project are multidirectional and must have an intent and a purpose, taking many forms and a variety of media. Thus, establishing and maintaining smooth and fluid working relationships across the organization is fundamental.

Relationship management focuses on project Stakeholders, to manage their respective expectations by ensuring that communication lines exist and are open. Regular appropriate distribution of project information and updates on project status, progress, and decisions will leverage stakeholder allies' support and alleviate the impact of project adversaries. Relationship management for the Project Manager is also key in project team management, by fulfilling the manager's role of leader, motivator, and delegator. Regular and punctual information exchanges with team members comprise stating group Objectives, explaining goals and standards, and keeping team members knowledgeable and aware of the context and content of the project.

4.3.1 The Project Manager's Relationship to Stakeholders

The Project Manager establishes different types of relationships as required to support the project's business and operational Objectives. These relationships are within the areas of the organization affected by the project and that provide resources to the project and to all external bodies (Figure 4.2).

The Project Manager should make an explicit effort to understand the full extent of the coverage of the project and cast wide to capture all potential Stakeholders who have some interest or level of influence that can impact the project. Stakeholders are not confined to the internal performing organization.

Several areas are of concern in relationship management, such as what expectations to focus on and to manage and with whom to forge the corresponding relationships. These areas are principally those relating to the results that meet the main enterprise Objectives, those that must be produced for operational goals, and the investment results achieved in capital development projects.

The capture of all Stakeholders to a detailed level is imperative. The model below is widely used as a framework to identify stakeholder involvement in the project, illustrating the internal/external organizational functional groups from which responsibility positions are subsequently identified and eventually a named individual is recognized, when it is possible.

Stakeholders can be grouped by organizational functional areas, both internal and external, or using the PESTLE categories (Figure 4.3).

When establishing stakeholder relationships, the Project Manager must be aware of and understand that Stakeholders are humans and have personal goals and political motivations. Their professional and personal interests will influence their decision-making. Their key motivations, however, are professional and their concerns will

Figure 4.2   The project stakeholder environment.

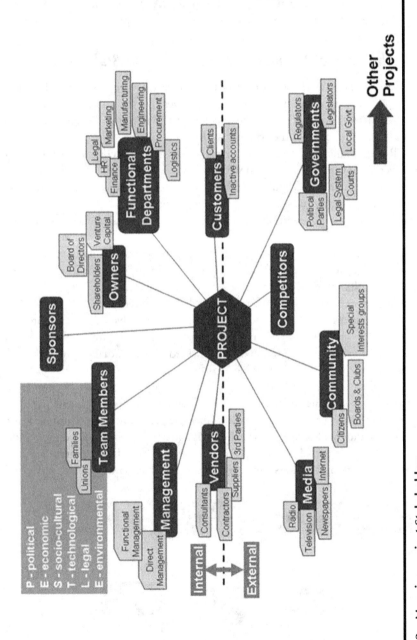

Figure 4.3   Mapping project Stakeholders.

differ at different stages of the projects. In some cases, for some Stakeholders, their interests will often go beyond the visible project goals.

4.3.2 Categorization of Stakeholders

4.3.2.1 The PESTLE Model

The PESTLE model is a structured tool widely used in the management sphere. Used in the project context, it allows the identification of Stakeholders by major areas of interest. The six categories of the model are political, economic, social, technological, legal, and environmental, where those Stakeholders that are identified will have particular expectations and interests. The model is to be specific to the project scope and to the Stakeholders that are mapped, and must only contain expectations that are relevant to the project. It is important to note that not all PESTLE areas of analysis may be pertinent to the project.

- **Political**: Stakeholders that will provide project expectations relative to local government, institutions, administrations, and financial policies
- **Economic**: Stakeholders that will provide the project with the relative enterprise business, organizational, and financial drivers, covering business cycles, financial/interest rates, and other Benefits Realization indicators
- **Social**: Stakeholders that will provide the project with the corresponding expectations covering internal and societal environments, such as demographics, lifestyle changes, attitudes to work and leisure, and levels of education
- **Technological**: Stakeholders that will provide the project with current and future technology and digitalization infrastructure needs and developments, and the eventual obsolescence of these
- **Legal**: Stakeholders that will provide the project with expectations as to the variety of laws that the project is required to comply with. These laws will encompass those which are national, international, employment, competition, environmental, health and safety, and those issued by federal/regional/local legislation
- **Environmental**: Stakeholders that will provide the project with expectations as to the environmental impact, energy consumption, waste disposal, etc.

4.3.2.2 The "Stakeholder Wheel"

For enterprise projects that are extensive across many levels of the organization, and are spread across many geographies, the Project Manager will deal with a large number of internal and external Stakeholders and must consider the multiple tiers of stakeholder involvement. Using the stakeholder "location circle" model illustrated in Figure 4.4, the Project Manager can structure and categorize the Stakeholders in expanding concentric circles, representing both functional area responsibility and geography.

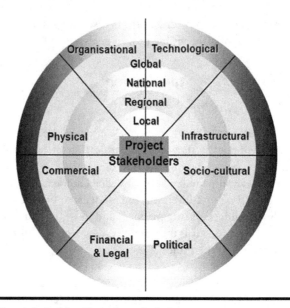

Figure 4.4   The stakeholder "location circle".

Primary Stakeholders are immediate communities of interest and can be marked inside the first circle. Secondary Stakeholders are the intermediaries in the process, and so on to the outer rim of the circles. The Project Manager will thus highlight many groups that do not think of themselves as Stakeholders.

4.3.2.3 The "3D Types" Classification

Stakeholders can also be categorized by their influence across the total timeframe of the ownership/operational cycle. A classification by "3D types", as illustrated in Figure 4.5, will assist the Project Manager to comprehend the sources of decision-making and influences that the project will need to manage.

Key Stakeholders who are behind the incentive of the project have the highest expectations as to the strategic and/or Operational Benefits and business value that the project must accomplish. These Stakeholders have a high-level view of the project and seek value delivery in the operational use of the project Deliverables. As illustrated in Figure 4.5 these Stakeholders are in the "drivers" category.

Figure 4.5   The project stakeholder 3-D types.

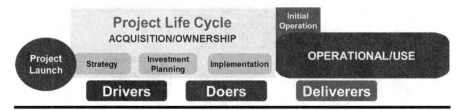

Figure 4.6   Project stakeholder 3D types across the Product Life Cycle.

Stakeholders who perform work during the project life cycle have a tactical view of their participation. These Stakeholders concentrate on the execution of project activities and tasks and their focus is on the production of the project Deliverables, be they internal or external to the project. These Stakeholders fall under the "doers" category.

Stakeholders who are responsible for exploiting the project's Deliverables (product or service) are part of the operational performing organization. Their expectations are to receive a usable product/service from which they can deliver the Operational Benefits and business value Objectives. These Stakeholders are the "deliverers" of the business value.

The "3-D" Stakeholders will have different concerns at different stages across the project life cycle, as illustrated in Figure 4.6.

4.4 Conducting a Stakeholder Analysis

Stakeholder Analysis is performed to identify and build a comprehensive list of the project's Stakeholders, to assess their interests, and to determine how they affect the project's viability and performance. The analysis establishes the goals and roles of different groups and formulates appropriate forms of engagement with these groups.

A Stakeholder Analysis is the first step in building the relationships needed for the success of a project. Importantly, it establishes the social environment in which Stakeholders operate and defines the approach to achieve this. A Stakeholder Analysis will (Figure 4.7):

- Identify and define the characteristics of Stakeholders
- Determine the interests of Stakeholders in relation to the problems that the project is seeking to address
- Discover conflicts of interest between Stakeholders
- Identify relations between Stakeholders that may enable "coalitions" of cooperation
- Assess the capacity of different Stakeholders to participate
- Establish the appropriate type of participation by different Stakeholders

Figure 4.7   The project Stakeholder Analysis process.

Stakeholder Analysis is a five-step iterative process:

- Identify project Stakeholders
- Identify Stakeholders' interests, impact level, and relative priority
- Assess Stakeholders for importance and influence
- Outline assumptions and risks
- Define stakeholder participation

4.4.1 Identify Project Stakeholders

Identification of the project's Stakeholders is a process of developing a list of those who are impacted by the project or will impact the project, positively or negatively. Core project team members and individuals from the organization familiar with the project's Deliverables and constraints, and the organization's structure and politics, should be involved in developing the list.

Identification is best done using a structured graphic similar to a mind-map and can be inspired by the PESTLE model. Stakeholders can be individuals, groups, communities, organizations, etc. Breaking stakeholder groups into smaller units will often assist in identifying important groups that may otherwise be overlooked. The list can be expanded following a brainstorming session or any other process commonly used in the organization.

Stakeholders can also be categorized by a useful graphical categorization, the stakeholder "functions wheel", as shown in Figure 4.8.

Once a comprehensive list of Stakeholders is established, the register shown in Figure 4.9 is initially populated.

The stakeholder register will subsequently be used to scale each stakeholder's interest, importance, and influence in relation to the project. Additionally, the stakeholder register will be the basis for developing the Stakeholder Communication Plan.

The process of identifying Stakeholders involves the participation of individuals who may have divergent views about the selected Stakeholders; often there is a need to challenge views and agreements made during this process.

Figure 4.8 The stakeholder "functions wheel".

STAKEHOLDER						SCALE			
Organization	Department	Title	Role in Project	Name or Function	3D Type	Interests	Importance	Influence	Priority
Internal	Board member	VP xxx	Sponsor	J. Top	Driver				
Internal	Project Department	Senior. PM	Project Manager	K. Nimble	Doer				
Internal	Manufacturing	Production Manager	Resource Manager	A. Giver	Deliverer				
Supplier X	Product Sales	Sales Manager	Project Leader	E. Deal	Doer				
Government Agency Z	Customs & Excise	Customs Agent	Customs Clearance	Customs Officer Lambda	Doer				
. . . .									

Figure 4.9 The stakeholder register.

4.4.2 Determining Interests, Importance, and Influence

The Project Manager is to schedule and hold meetings and interviews with the Stakeholders listed during the first step of the analysis. Discussions should be open and frank to draw out the key interests of each stakeholder in relation to the problems that the project is seeking to address.

For best results, the process of identification should be conducted as a series of facilitated workshops. Through this, knowledge about Stakeholders, their interests, power, and influence can be uncovered and documented.

Individually, the team members will benefit from exposure to new ways of understanding relationship management and will learn about the characteristics, leadership and management styles, and expectations of the project's key Stakeholders.

From the stakeholder categories, or other grouping techniques, the Project Manager can build the context of the meetings by structuring questions that concern the target category. For example, business managers are more concerned about

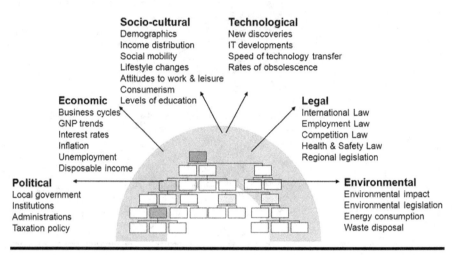

Figure 4.10 The PESTLE interest spectrum.

financial and organizational issues, while technology experts will focus more on infrastructure and processing issues. The PESTLE interest spectrum shown in Figure 4.10 can be used as a guideline for structured questions:

The Project Manager should conduct discussions by inviting the Stakeholders to express themselves on how they perceive the project and what guidance they could offer.

Key questions could include:

- What are the likely expectations of the project by the stakeholder?
- What role would the stakeholder wish to hold in the project?
- What benefits are likely to be there for the stakeholder?
- What negative impacts may there be for the stakeholder?
- What resources is the stakeholder likely to commit (or avoid committing) to the project?
- What other interests do the stakeholder have that may conflict with the project?
- How does the stakeholder regard others on the list?

The Stakeholders' interests are recorded in the stakeholder register on a scale of 1 to 5 (Figure 4.11).

4.4.3 Assessing Stakeholders for Importance and Influence

This step assesses the importance and influence of each stakeholder on the project. A scale from 1 to 5 can be used for each assessment, where 5 is the most important or most influential.

Importance is scaled to represent how much the stakeholder's issues, expectations, needs, and interests are related to the aims of the project. If the highest-scaled important Stakeholders are not involved or assisted, then the project will be on track for failure.

STAKEHOLDER						
Organization	Department	Title	Role in Project	Name or Function	3D Type	Interests
Internal	Board member	VP xxx	Sponsor	J. Top	Driver	
Internal	Project Department	Senior. PM	Project Manager	K. Nimble	Doer	
Internal	Manufacturing	Production Manager	Resource Manager	A. Giver	Deliverer	
Supplier X	Product Sales	Sales Manager	Project Leader	E. Deal	Doer	
Government Agency Z	Customs & Excise	Customs Agent	Customs Clearance	Customs Officer Lambda	Doer	
. . . .						

Figure 4.11   Recording stakeholder interests.

STAKEHOLDER							SCALE		
Organization	Department	Title	Role in Project	Name or Function	3D Type	Interests	Importance	Influence	Priority
Internal	Board member	VP xxx	Sponsor	J. Top	Driver				
Internal	Project Department	Senior. PM	Project Manager	K. Nimble	Doer				
Internal	Manufacturing	Production Manager	Resource Manager	A. Giver	Deliverer				
Supplier X	Product Sales	Sales Manager	Project Leader	E. Deal	Doer				
Government Agency Z	Customs & Excise	Customs Agent	Customs Clearance	Customs Officer Lambda	Doer				
. . . .									

Figure 4.12   Scaling stakeholder interests, importance, and influence.

Influence is scaled to represent both the formal and the informal power held by the stakeholder. Hierarchical positions such as project sponsor, management or project steering committees, operational managers, and primary users would score high. This would also be the same for informal influencers such as a project champion, a subject-matter expert, or a key external provider.

The scores are reviewed, agreed upon, and recorded by the project team. A first-order priority and ranking is then made by calculating the product of the interests, importance, and influence scores (Figure 4.12).

4.4.4 *Stakeholder Expectations and Strategic Goals*

The exercise of analyzing and mapping the stakeholder base will improve the Project Manager's insight into the Stakeholders and their drivers. The interest/importance/

influence mapping will indicate how each individual stakeholder relates to the project's goals and requirements.

The Project Manager and the project team can now focus on the assumptions relating to both the project's initial scope and the stakeholder's expected consistency throughout the development life cycle. For medium- or long-duration projects and/ or complex internal/external inter-relationships, changes in the composition of important Stakeholders and their expectations are to be anticipated and the underlying assumptions recorded.

The risks assessed in the Stakeholder Analysis refer more to the nature of the Stakeholders, the level of their interests, the viability of their expectations, and how they rank in priority. All these key factors will place uncertainty on the stability of the project scope and thus have an impact on the project's success. For example, low-interest Stakeholders who hold high importance and influence can be resistant to the project's goals and create obstacles outside of the control of the Project Manager.

4.4.5 Define Stakeholder Participation

This step closes the Stakeholder Analysis cycle, with the Project Manager and the project team completing an assessment of the capacity of different Stakeholders to participate and determining their appropriate type of involvement at successive stages of the project cycle.

The project should plan strategies for approaching and involving each individual or group of Stakeholders. Special attention should be given to reluctant Stakeholders, while a specific monitoring mechanism should be instituted for those Stakeholders who may change their level of involvement as the project is performed. When the stakeholder is a group rather than an individual, the Project Manager will need to decide whether all members of the group participate or only selected representatives of the group.

The Project Manager can establish a first scheme of involvement using the importance/interest grid as follows, other combinations between interests, importance, and influence can also be established (Figure 4.13).

Stakeholders are positioned according to their respective importance and interest scores, as determined previously in Section 4.3.2.1.

■ High importance and high interest: To fully *engage* the Stakeholders and make the greatest efforts to satisfy them
■ High importance and low interest: Place enough effort to *satisfy* the Stakeholders, with no excess to avoid weariness of the message
■ Low importance and high interest: *Inform* the Stakeholders adequately, to ensure that no major issues arise. To utilize these Stakeholders sensibly as often they provide assistance and help with the details of the project
■ Low importance and low interest: To *monitor*, with no excessive communication

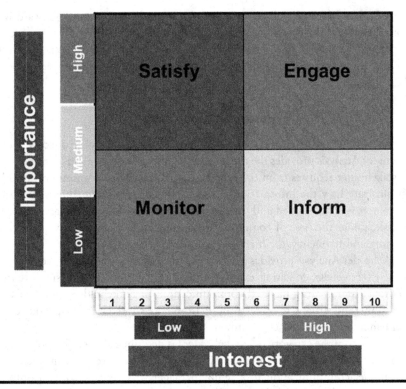

Figure 4.13   Mapping stakeholder participation.

The matrix maps the Stakeholders according to their ability to influence the project's success or failure. The Project Manager must seek to appreciate and recognize the stakeholder interests and also establish an involvement status according to their "interest duration" throughout the project, which could be continuous, frequent, intermittent, sporadic, or one-off.

The Project Manager and the core team should then complete the Stakeholder Analysis by defining for each stakeholder their appropriate involvement, their level of support, and their receptiveness to messages about the project. The team can then proceed to draft the Stakeholder Communication Plan.

4.5 The Stakeholder's Project Communication Environment

The fluidity of communication lines between the Project Manager and the Stakeholders is key to understanding and maintaining expectations. A conducive communication environment will forge a relationship based on common understanding and alignment between parties and create a collaboration platform focused on the realization of a result-oriented conversation goal.

The Project Manager will communicate in a "360" environment: upward with senior managers of the organization; sideward with line managers and peers, such as other Project Managers; outward with contributing project Stakeholders, including end users, line managers, shareholders, unions, suppliers, and government bodies; and downward with core project team members.

4.5.1 Stakeholder Communication Plan

Frequent and comprehensive communication is a key project success factor. Stakeholder Analysis provides the expectations of the stakeholder and identifies what each stakeholder requires from the project. The Stakeholder Communication Plan must illustrate how the information needs of all project team members and other Stakeholders will be satisfied and verified with a feedback loop. The plan defines for each stakeholder the type of communication most likely to be effective, specifying the timing and frequency, the medium, and the expected outcomes.

Stakeholder Analysis provides a key to the communication plan by seeking multilateral effectiveness, soliciting explicit actions to gain and sustain support from advocates, disseminating targeted communication for Stakeholders with "volatile" expectations, and striving to achieve buy-in from critics and recalcitrant Stakeholders and enhancing interest and commitment (Figure 4.14).

Derived from the category of a stakeholder (as described above), this will provide the approach to how the communication plan should be developed and delivered. For example, communication to a senior manager will need to contain only the information in the format necessary to provide management with essential data about the project, while communication with team members will need more details and use a different language.

The Project Manager will use the Stakeholder Communication Plan to engage in networking with influential Stakeholders, to gain and sustain support from advocates, and to achieve buy-in from critics. The Project Manager must strive to communicate in the stakeholder's "language".

Stakeholder	Objectives	Type of Communication	Medium	Frequency and Schedule	Distribution	Outcome
XXXX	XXXX	XXXXXXXX	XXXX	XXXX	XXXX	XXXX

Figure 4.14   The project Stakeholder Communication Plan register.

For effectiveness, the Project Manager is to communicate according to an agreed-upon frequency while ensuring "readability" by focusing on the core topic and making it compelling and interesting. The use of a vocabulary that is compatible with the audience is essential and the communication must avoid buzzwords, three-letter acronyms where possible, and esoteric terminology. E-mail communication is to be concise and cover one topic at a time, while avoiding emotional and dramatic phrases so as not to lose credibility.

The effectiveness of the communications plan is to be frequently assessed and results integrated into lessons learned.

During the project life cycle and at an appropriate time, frequent meetings with Stakeholders offer crucial platforms to manage expectations, to monitor their engagement and relay information, and to maintain constant communication channels.

The results of the Stakeholder Communication Plan are integrated into the project communication plan and the project schedule. During project implementation, they are monitored through team meetings and regular reports.

The Project Manager cannot expect to align all Stakeholders all the time. The project's stakeholder community changes as Stakeholders move within the organization or leave it. Consequently, new Stakeholders have to be included and current Stakeholders will experience changes in their relative importance to the project, as well as their power and influence. As the project proceeds through its life cycle phases, different Stakeholders may have more or less of an impact, and as a corollary, their levels of importance and communication requirements will change.

The key issues to address are:

- Is the Stakeholder Communication Plan as effective as it should be?
- What additional actions should be taken to gain more support from the project advocates?
- What are the strategies to implement to convince critics?
- Does the stakeholder importance/interest matrix reflect the current state?

The Project Manager will also need to establish a monitoring scheme to capture:

- Who influences the Stakeholders' opinions generally?
- Who else might be influenced by their opinions?
- Do influencers therefore become important Stakeholders?
- If the influencers are not likely to be positive, what will be required for them to support the project?
- If there is little possibility or probability of gaining their support, how will their opposition be managed?
- What is their current opinion of the Project Manager?
- Who influences their opinion of the Project Manager?

Because of the dynamics of the stakeholder community, the effectiveness of the Stakeholder Analysis and the Stakeholder Communication Plan will suffer if not adapted to the evolving situation. The Project Manager must therefore plan for and prepare the project team to repeat the Stakeholder Management process in its totality, or in part, many times as the project progresses through its life cycle phases or as the stakeholder community changes.

4.5.2 Establishing the Project's Stakeholder Relationship Map

The project's Stakeholder Relationship Map illustrates the organizational context in terms of the connections that exist among and between the internal and external Stakeholders. The map indicates the functions performed and the flux from these functions between entities. Relationship arrows represent formal information exchanges and known informal exchanges.

For projects that address an exceptionally large number of Stakeholders from both the internal organization and the external entities, a valuable tool to enhance the understanding of the multilevel and multilateral communication lines is a Stakeholder Relationship Map as illustrated in Figure 4.15. The map, if useful, can also be used when projects concern a handful of Stakeholders.

The purpose of the relationship map goes beyond depicting the spread of Stakeholders and their bilateral and multilateral communication lines. The map illustrates where and how Stakeholders may communicate "out-of-sync" with the project's communication plan.

From an analysis of the relationship map, and in conjunction with the previously developed Stakeholder Communication Plan, the Project Manager can establish specific communication contents for specific stakeholder groups at specific points in time and ensure that the communication method is pertinent to the audience and the topic.

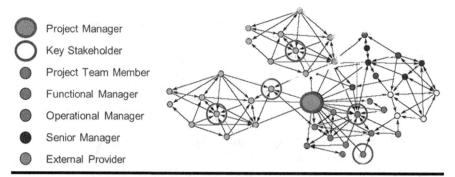

Figure 4.15   The project's Stakeholder Relationship Map.

The Project Manager cannot possibly be present at every stakeholder-to-stakeholder exchange, whereas the map can provide areas the Project Manager, when absent, may rely on certain key Stakeholders to act as champions or promoters of the project.

- Direct relationships between the project and Stakeholders are with:
  - Core team members
  - Key Stakeholders and sponsor
  - Steering group
  - Upper management (on the direct report)
  - Other Project Managers (on the interface)
  - Functional managers (as resource and performance providers)
  - Operational managers (as users)
  - External providers (on project performance)

- Indirect relationships between the project and Stakeholders are with:
  - Upper management
  - Peers
  - Project team members (reporting to team leaders)
  - Team members of interfaced projects
  - Functional and shared services managers
  - External government and institutions

4.5.3 The Informal Network

The internal workings of an enterprise are a daily gathering of a constituted community of individuals issued from a variety of origins, both geographic and cultural. The institutionalized organizational chart only indicates the vertical and lateral lines of formal communication. Every community, and thus an enterprise, will also function along an informal network not described in a clear and visible manner.

Informal networks exist between those who have common interests, shared values and philosophies, family ties, friendly ties (school, clubs, community work), professional and academic groups, organizational and career ambitions, etc.

Informal networks aim to achieve a mutual and reciprocal exchange of information and allow the pursuit of personal goals. Members of an informal network seek to help each other obtain personal, business, and career advantages.

The Project Manager must be skilled to work and communicate with the informal network, to determine where on the relationship map these exist, and be proactive in initiating conversation with the concerned individuals.

By communicating on a regular basis with each stakeholder on the network, both formal and informal, the Project Manager can demonstrate personal and professional skills using a business vocabulary, and establish increased credibility.

Possessing the ability to ask the pertinent questions and not hesitating to ask for direction and opinions will also help create stronger connections.

4.6 The Project Manager's Communication with Stakeholders

The Project Manager's responsibility is to deliver the project to expectations, and authority must be a formal and clear assignment from the sponsorship group. The authority must be compatible with the decision-making required, and the level of authority is relative to the autonomy assigned to the Project Manager within the organizational structure. Key project Stakeholders must be informed of the Project Manager's authority level.

As noted above, the Project Manager expends more than 75% of the project management time in "communication mode". It is imperative therefore that the Project Manager possesses highly developed human and soft skills to relate effectively with all individuals associated with the project.

The Project Manager has neither all the competencies and skills to perform all the project work nor the material time to do so. A major principle to be retained is:

> A person responsible for a group of individuals must understand that Objectives given to be realized can only be accomplished through the efforts of others.

The Project Manager will often have no hierarchical power and has to place the focus on the achievement of project goals, which will require compatibility of understanding from all Stakeholders. It is therefore of utmost importance that the Project Manager create a working environment with all Stakeholders that is inspiring, motivating, and rewarding.

The frequency, medium, and level of communication with all Stakeholders will vary depending on their role and position with regard to the project.

4.6.1 Sponsorship Group Communication Interface

The sponsorship group's role is to foster support for the project within the senior management levels of the organization and key operational managers. Most successful projects have sponsors who understand the importance of their involvement as it is a fundamental activity and plays a critical role in the eventual success or failure of the project.

It is essential to build bilateral, fluid, and open communication with the sponsorship group in order for the latter to demonstrate and promote both public and private support to the project. Communication with the sponsorship group and upper-level management is to be succinct and highlight issues, risks, and exceptions.

Face-to-face reviews, often for a short duration, are to be conducted at the summary levels of information; however, the Project Manager must be in a position to provide details when requested.

4.6.2 Internal Organization Communication Interface

Lateral communication with peers from the internal organization will revolve around meetings for the negotiation of resources and agreement on the plans for the schedule of project work. Operational functional managers are the principal source of internal resources and have to balance their own operational performance needs with the project assignments requested. The Project Manager's communication is to be focused on the organizational requirements and the success factors to achieve them, so as to shift the communication exchange away from a personal interchange to one at an enterprise level.

4.6.3 Project Team Communication Interface

Communication with the project team members is primordial to providing direction, project goals and Objectives, and the constrained timeframe for execution. Project plans, schedules, resource needs, risks, and funding have to be readily accessible to all participants on intranet platforms or the like. The Project Manager must also allocate a high proportion of time to verbal and face-to-face exchanges. This goes beyond just project meetings on progress and status, scheduled activities, and pending work to perform. When and where possible, the Project Manager should apply the MBWA principle – management by walking around – to be seen and to be accessible to project team members.

4.6.4 Contractors and Suppliers Communication Interface

Projects utilize contractors and suppliers because their services, products, knowledge, and expertise are required for success. These Stakeholders play a critical role in the project.

Organizations seek external services when they do not know how to do the work, do not have the present possibilities for doing it, or just do not want to do it. Thus, it is clear that as organizations seek help, it is therefore comprehensible that the Project Manager must engage in communication that is appreciative.

The Project Manager must consider all contractors and suppliers as team members and not as outsiders, while maintaining professional and contractual relationships with them. Key preferred contractors and suppliers may be invited to integrate the project planning and decision-making process.

Facilitating a communication environment where both parties work as partners, where and when possible, will enhance collaboration and mutual decision-making when faced with issues. The Project Manager must not put unnecessary pressure on contractors and suppliers, as these Stakeholders are essential to the project.

4.7 People and Communication Skills for the Project Manager

Projects are traditionally considered to be an essentially technical process.

The Project Manager is to principally focus on project Objectives and goals, scope content definition and specification, front-end engineering, activity planning and scheduling, resources identification and funding requirements, project monitoring and reporting, and overseeing the diverse technical subject-matter expertise to be employed. Although this nonexhaustive list is very acceptable, none of the items cited will be performed to an acceptable quality standard and level without specific and special attention to people.

Important to project management are people, the communication skills that a Project Manager possesses, and the competencies in leadership, motivation, and team dynamics displayed and utilized. The Project Manager's communication aims are to create awareness, understanding, engagement, involvement, and commitment.

The key communication skills for Project Managers include verbal and nonverbal communication, active listening, open-mindedness, friendliness, trust, and respect, and the ability to establish collaboration with Stakeholders.

The Project Manager's talent to encourage project team members, and to fully utilize and develop their abilities, inspires team members, enhances confidence, and results in better project performance.

Engaging with Stakeholders at all levels in an open, friendly, and professional manner, using the panoply of communications skills, enables the Project Manager to disseminate information to Stakeholders and to generate and maintain stakeholder commitment. Addressing resistance, objections, and conflicts further tests the Project Manager's communications skills and ability to resolve project issues.

4.7.1 Communication Distribution – Verbal and NonVerbal

A large percentage of our face-to-face communication displays emotions, which are influenced 50 percent by nonverbal cues, 40 percent by the tone of voice, and 10 percent by the verbal/spoken word. With the power of the spoken word, language paints our message and is the representation of our experiences, and by using a vocabulary, we generalize, filter, and at times, distort the message.

Our sensory input is correlated with our personal experience, where we extract information from the world that surrounds us, depicting our own representative models of the world and filtering what we see and hear to conform to our needs.

When we transmit information, be it visual, auditory, or tactile, to be coherent we need to consider the ensemble of our body language, tone of voice, and choice of words. All signals are significant and must be taken in context. We should be aware of our body language attitudes: open hands, touching the chin, frowning, startled eyes, finger raised, fist clutched, jaw muscles tense, fidgeting, playing with hair, and so on.

Figure 4.16 is a freely adapted illustration based on the Shannon–Weaver communication model for a bilateral face-to-face exchange.

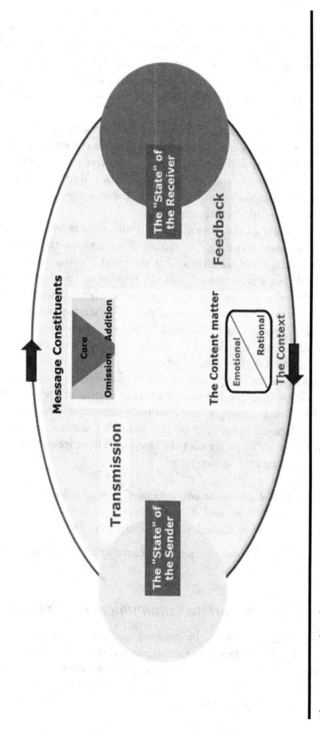

Figure 4.16 The Shannon–Weaver communication model.

The "sender" formulates a message to transmit in a certain "state". The message has core constituents, which may not be clear and obvious, with additions and omissions. The message may be enveloped in a mix of emotional and rational "states". The message thus has been "coded". As the message is transmitted to the receiver, it will pass through a zone of "noise". This may be physical, surrounding sounds, for example, to softness of speech and to even a specific language vocabulary.

The "receiver" is the recipient of the message and is in a "state". The message has to be decoded, including by eliminating where possible the effects of the "noise", as the receiver attempts to decipher the message constituents and unravel its emotional/rational characteristics. The "receiver" then sends a "feedback" (consciously or not) according to the receiver's comprehension of the message and follows the same path as that used by the original "sender".

Following the study of the human nervous and sensory system, science suggests that there is approximately a half-second delay between the instant that the senses receive a stimulus and the instant that the mind is conscious of a sensation. However, the body has a reflex system that can respond in less than one-tenth of a second, before the mind is conscious of the stimulus. This half-second delay seems to be the time required for processing and compressing sensory input.

Thus, we "feel" four-tenths of a second before we "know", and our nonverbal body language has already signaled to the receiver how we "feel" about the received message. Just consider a slightly raised eyebrow.

It can therefore be easily appreciated that in face-to-face communication the number of pitfalls are multiple and lead to a variety of miscomprehensions, misunderstandings, emotional exchanges, and eventually physical altercations.

The Project Manager must practice and hone the necessary communication exchange skills to convey messages with the intended Objectives in mind.

The author recognizes and has freely used:

A Synthetic Model representation, expanded from the Shannon-Weaver Model, Warren Weaver and Claude Elwood Shannon (1963). The Mathematical Theory of Communication. University of Illinois Press.

and "Behavior studies conducted by Dr Albert Mehrabian"

4.7.2 Using the Appropriate Communication Medium

The major groups that constitute the medium used in communication are oral, written, and visual. The Project Manager disposes thus of a large palette of communication vehicles. It is important to employ the most appropriate medium and prepare and structure the message contents to ensure that misunderstanding and miscomprehensions are kept to a minimum.

4.7.2.1 Oral Communication

Oral communication covers meetings with Stakeholders involving the project and team members, sponsorship group and management, functional managers, operational personnel, external providers, and government bodies. These meetings may be either with the physical presence of participants or as virtual online sessions, or a mix of the two. Thus, body-language nonverbal communication will be at the forefront, unless virtual meetings are oral only and video transmissions are switched off. Oral communication also covers the use of a telephone or electronic connection. This communication can be exchanged between two or more individuals, including audio conferences, or can be exchanges posted as messages to be listened to at a later stage. Here, only the verbal attribute of the exchange is accessible.

4.7.2.2 Written Communication

Written communication covers a vast range of verbal and grammatical formulations. Some examples would be project reports of all types, letters and emails exchanged with internal and external Stakeholders, social media platform messages for inter-project reporting, project newsletters and press releases, enterprise messages on information boards, etc. While it is not expected that all project participants will be majors in literature, it is important to prepare and formulate what is to be conveyed and to review the message prior to its dispatch. Too many hasty messages using electronic means prove to be the downfall of many an individual. The written word remains, as well as its interpretation by different people.

4.7.2.3 Visual Communication

The visual communication medium (excluding virtual online meetings) is a mix of oral and written forms. This type of communication covers paper, electronic, and video and is typically an outgoing transmission. Project methodologies, instruction manuals, design schematics, operational processes, contracts, etc., would typically be on paper only or on paper and in electronic formats. Electronic formats would be for presentations to a group and can also be distributed on paper or recorded as a video. Video communication may well be used specifically to develop educational contents, instructions on use of equipment, etc. There will certainly also be an oral and written content to these.

4.7.3 Developing Rapport with Open Communication

Rapport is built upon open communication, which is a positive attitude of interest in the other person and exploration of their interests. Rapport is formed when two or more people feel that they are in sync or harmony because they feel similar or relate well to each other, and there is animosity. Rapport includes three behavioral components: mutual attention, mutual positivity, and coordination.

The Project Manager must strive to build rapport and relationships across the stakeholder community. This is done by primarily engaging in being honest, sincere, ethical, and professional and having integrity. By communicating clearly and frequently, the Project Manager can seek commonalities with others and bond and connect with people.

Rapport building is an explicit set of actions that the Project Manager must initiate to establish open communication with the stakeholder community.

4.7.4 Active Listening and Empathy

Active listening and empathy imply listening attentively to a speaker, understanding what they're saying, responding and reflecting on what's being said, and retaining the information for later. No judgment is made. This keeps both listener and speaker actively engaged in the conversation.

Active listening involves going beyond simply hearing the words that another person speaks but also seeking to understand the meaning and intent behind them. It requires being an active participant in the communication process. The importance is to make the other person confident that they are heard and valued.

Paying attention to, and recognizing, nonverbal cues will allow one to understand the message and what the other person is aiming to say. Multiple verbal cues have to be considered together, so as not to misinterpret the body language: yawning does not mean a person is bored, but maybe rather tired due to overwork. If the person is talking fast, for instance, this could be a sign that they are nervous or anxious, or just excited depending on the words used. If they talk slowly, they may be trying to carefully choose their words or maybe just tired or they may be of a reserved nature.

The basic rule for active listening and empathy is foremost to take up the responsibility for the success of the communication. Be aware of the exchange of verbal and nonverbal signals and reduce emotions from the message when these are perceived. Perceive the emotions of the other person and position yourself in their situation. Avoid distractions, concentrate, and be alert. Identify the sender's goal and seek the principal themes of the communication. Do not arrive at the other person's conclusion before listening to the totality of the message. Acquiesce the other person's message and nonverbal signals by synchronizing your tone of voice and body posture to theirs. Use words to maintain the rhythm of the conversation and when possible rephrase the person's message to seek the common agreement of its content.

In face-to-face physical communication, your body posture is of great importance and you should be positioned facing the other person without any appearance of discomfort or annoyance. In several cultures, eye contact is key to emphasize your readiness and acceptance to communicate.

Gauge the impact of your communication so as not to create any barriers to make the other person feel threatened and enter into a defensive communication mode. Be descriptive, problem-oriented, spontaneous, empathetic, equal, and reserved. Barriers are created when you take an attitude of being evaluative, controlling, strategic, neutral, superior, or certain.

4.7.5 Communication Challenges and Breakdowns

The major causes for communication breakdown are numerous: we do not "speak" the same language or have the same interests. In projects, the major causes are unaligned expectations, unclear agenda, unclear direction and goals, unclear priorities, a lack of leadership, and unclear roles and responsibilities.

Furthermore, communication breaks down due to other causes such as the following non-exhaustive list:

- Language differences
- Cultural differences
- Lack of cultural awareness
- Different goals
- Different levels of knowledge
- Information "hiding"
- Bad communication skills
- Bad intentions
- Lack of respect
- Lack of structure
- Lack of trust
- Climate of fear
- Lack of energy, sleep, and food

Communication breakdowns also exist due to a lack of understanding of emotions and personality traits. Figure 4.17 has been adapted from "Managing People" by

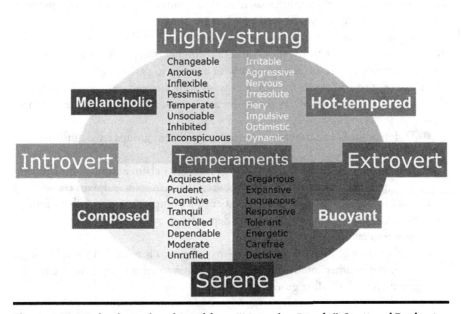

Figure 4.17 Behavior traits adapted from "Managing People", Scott and Rochester.

Scott and Rochester and relates to behavior traits that can be manifested by both parties during an interchange. None of these is "bad" and the Project Manager must recognize shifts in behavior to maintain communication on an even keel.

4.8 Sources of Conflicts in Projects and Stakeholder Management

This encompasses a wide range of conflicts that people experience in organizations, such as incompatibility of goals, differences over interpretations of facts, and disagreements based on behavioral expectations.

Reputable dictionaries offer a range of definitions as to what conflict is and/or where it originates from:

- A struggle or clash between opposing forces; battle
- A state of opposition between ideas, interests, etc.; disagreement or controversy
- A clash, as between two appointments made for the same time
- Psychological opposition between two simultaneous but incompatible wishes or drives, sometimes leading to a state of emotional tension and thought to be responsible for neuroses
- To come into opposition, clash
- To fight

A traditional view of conflicts in projects holds the belief that all conflict is harmful and must be avoided from causes such as poor communication, lack of openness, failure to respond to employee needs, and so on.

The human relations view of conflict holds the belief that conflict is a natural and inevitable outcome in any group and that conflict is not only a positive force in a group but that it is absolutely necessary for a group to perform effectively.

There exist several causes for conflict throughout the management of a project; a short selection is given below, as the causes are too numerous to cite in this section:

- Different goals and timeframe horizons – different groups have different goals and focus
- Overlapping authority – Two or more managers claim authority for the same activities, which leads to conflict between the managers and workers
- Task interdependencies – one member of a group or a group fails to finish a task that another member or group depends on, causing the waiting person or group to fall behind
- Different evaluation or reward systems – a group is rewarded for achieving a goal, but another interdependent group is rewarded for achieving a goal that conflicts with the first group

- Scarce resources – managers can come into conflict over the allocation of scarce resources
- Status inconsistencies – some individuals and groups have a higher organizational status than others, leading to conflict with lower-status groups

4.8.1 Types of Conflicts in Projects

The major types of conflict are functional and dysfunctional which provoke conflicts within the group and lead to a flow of related subordinated conflicts.

4.8.1.1 Functional Conflict

Functional conflict challenges the status quo and provides a medium through which problems can be aired and tensions released and fosters an environment of self-evaluation and promotes reassessment of group goals and activities.

Functional conflict supports the goals of the group and improves its performance by increasing information and ideas, encouraging innovative thinking, promoting different points of view, and reducing stagnation.

4.8.1.2 Dysfunctional Conflict

Dysfunctional conflict within an organization is motivated by the egos of employees with competing ambitions. It hinders group performance while reducing trust, creating tensions and stress, and leading to poor decision-making. And excessive focus on the conflict stifles the few new ideas that can be generated.

4.8.1.3 Organizational Process Conflict

This refers to conflicts over how work gets done between managers and subordinates, between departments, and at functional interfaces.

4.8.1.4 Project Task Conflict

This refers to conflicts over work content and goals at the start of group development and among team members as to ways of reaching the goals

4.8.1.5 Individual Relationship Conflict

This refers to interpersonal relationship conflicts between two or more people over a difference in the ways something should be done. An important cultural element exists which can be deemed to be a threat to an individual's value system and considered to be unfair treatment.

In the management of projects, this type of conflict can be interpersonal between individuals due to differences in their goals or values or intragroup, within a group or team, extending to conflicts between two or more teams, groups, or departments. Unchecked, this type of conflict may become interorganizational, rising and spreading across organizations.

The focus of the Project Manager is to engage in being cooperative and attempting to satisfy the other party's concerns, while being assertive and attempting to satisfy one's own concerns. Functional conflicts will require resolution by compromise or collaboration.

4.8.2 The Conflict Characteristics

Conflicts can be difficult to control once they have begun as the trend is toward escalation and polarization. When conflict escalates to the point of being out of control, it almost always yields negative results.

The Project Manager must recognize when conflicts develop into an uncontrollable state by intervening to separate the people from the problem and to focus on interests and not positions. Inventing or proposing options for mutual gain and seeking objective criteria may appease the conflict. Conflicting situations may be addressed by an understanding of the difference between opposition and cognition and personalization.

4.8.2.1 Opposition

Conflict opposition in projects is often due to structural conditions, where opposition may exist to the goals to be achieved; the nature of work; and the recognition and reward system and leadership styles may not be appropriate. A lack or insufficient competence in the Project Manager's communication skills will lead to misunderstandings, and at times, it is just a matter of semantics. Personal variables between individuals because of different personality types or values systems will generate opposition in their views on how to approach a problem or an issue. The opposition will block all project progress and may sour the relationships even more.

4.8.2.2 Cognition and Personalization

The Project Manager must bring together the parties so that they place their differences aside and decide what the conflict is and the reasons it exists – this is an important cognitive step. Time and space must be allowed for individuals to think and rationalize on the conflict.

Perceived conflict is the awareness by one or more parties of the core reasons for the conflict and signifies the existence of conditions that create opportunities to address the conflict.

Felt conflict is the emotional involvement of the parties in the conflict. Positive emotions help in finding solutions to resolve the conflict, while negative emotions may create a stalemate due to anxiety, frustration, and animosity.

4.9 Dealing with Conflicts

When addressing conflicts, two major forces drive the intentions of individuals: assertiveness and cooperation. These intentions are initially motivated by what the conflict issue is, how it is felt by the individual, and the behavioral traits of the person's core personality type. For example, a Project Manager who wishes to resolve a resource scarcity problem may initiate with a cooperative intention, while that same Project Manager may initiate with an assertive attitude when demanding a result. It may well be that the core personality type of the Project Manager is not that demonstrated by the intention.

Intentions from different parties may well evolve and change as the conflict is addressed. An individual's intention will face that of the interlocutor's, and when at odds, may aggravate the original conflict.

4.9.1 Key Model for Conflict Management

The well-known model in Figure 4.18 identifies five intentions to address conflict situations.

The model identifies the intentions through which both parties will traverse as the conflict resolution is addressed. As stated above, each party may well initiate with a specific attitude, be it assertive or cooperative, and then may either stay there or evolve as the conflict resolution progresses.

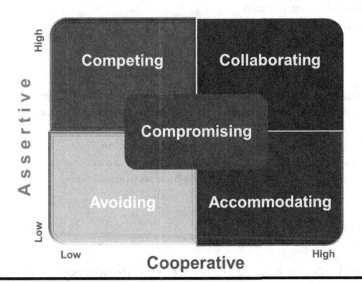

Figure 4.18   Assertiveness and cooperation model for conflict management.

Source: K. Thomas, "Conflict and Negotiation Processes in Organizations," in M.D. Dunnette and L.M. Hough (eds.), Handbook of Industrial and Organizational Psychology, 2nd ed., vol. 3 (Palo Alto, CA: Consulting Psychologists Press, 1992), p. 668. With permission.

4.9.1.1 Competing (Assertive and Uncooperative)

Competing refers to a desire to satisfy one's interests, regardless of the impact on the other party to the conflict. Each party tries to maximize its own gain and has little interest in understanding the other's position. Best to use when:

- Quick, decisive action is vital: in emergencies; for important issues
- Unpopular actions need implementing (in cost cutting, enforcing unpopular rules, and discipline)
- On issues vital to the organization's welfare
- One party knows it is right

4.9.1.2 Collaborating (Assertive and Cooperative)

A situation in which the parties each desire to satisfy fully the concerns of all parties and handle the conflict without making concessions by imagining a new way to resolve differences that are more beneficial to both. This is often called a win-win situation. Best to use:

- To find an integrative solution for an issue that cannot be compromised
- To combine insight with different viewpoints
- To gain commitment
- To repair a relationship
- Learning

4.9.1.3 Avoiding (Unassertive and Uncooperative)

The desire to withdraw from or suppress a conflict, where the two parties try to ignore the problem and do nothing to resolve the disagreement. Best to use when:

- Information is required before immediate decision
- Issue is insignificant or more important issues are to be addressed
- Potential disruption outweighs the benefits of resolution
- Others can resolve the conflict effectively
- Issue seems indicative of other issues
- To regain composure

4.9.1.4 Accommodating (Unassertive and Cooperative)

The willingness of one party in a conflict to place the opponent's interests above its own interests, simply giving in to the other party. Best to use when:

- Harmony and stability are especially important
- Issues are more important to others than to yourself and to maintain cooperation

- To allow a better position to be heard
- To build social credits for later issues
- To minimize loss when you are wrong, outmatched, and losing

4.9.1.5 Compromising

A situation in which each party to a conflict is willing to renounce something, and each party seeks to accomplish a goal and is willing to engage in finding a middle ground to reach a reasonable solution. Best to use when:

- Goals are important but more assertive approaches may create a potential disruption
- To reach provisional agreement over complex issues
- To arrive at a decision under time pressure
- Opponents with equal power are seeking mutually exclusive goals
- To fall back when collaboration or competition is unsuccessful

4.10 The Principles of Influence

Influence is the capacity to have an effect on the behavior, character, or development of someone or something.

- When Person A has the ability to influence the behavior of Person B, Person B acts in accordance with Person A's wishes
- When Person A possesses something that Person B requires, Person B's relationship with Person A is based on dependency

Social influence occurs when an individual's thoughts and actions are affected by other people. The forms of social influence are varied and are characterized by conformity, socialization, persuasion, peer pressure, and obedience.

Identification is when people are influenced by someone who is liked and respected and internalization is when people accept a belief or behavior and agree both publicly and privately.

Compliance, on the other hand, is when people appear to agree with others but actually keep their dissenting opinions private.

4.10.1 The Influence Model – Foundation and Structure

Self-esteem is a core personal value, and the foundational components for influence are driven by the perception of the individual of a given situation (Figure 4.19):

- Worth – "what's in it for me?"
- Likelihood – "is this credible?"
- Effort – "how much do we have to put into this?"

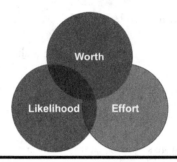

Figure 4.19　The influence model – foundation and structure.

The more the three components overlap, the more the individual is prone to the influence exercised on that person. However, when the more the gaps between the components, the less the individual can be influenced.

Associated with these central three components, the following factors also condition how effective the exercise of influence will be:

- Capability – "can I actually do this?"
- Opportunity – "what's the probability of success?"
- Threat – "where can I fail?"

The Project Manager must ensure that the foundation and associated factors are addressed to have any possibility of influence, as people align their personal values to the needs of the task as they perceive it: security, autonomy, recognition, power, success, financial reward, challenge, task variety, etc. When the Project Manager strives to satisfy the majority of the individual's perception, the resulting behavior will determine the effectiveness of the influence when the individual expresses:

- Commitment – signaling the individual is engaged
- Acknowledgement – signaling the individual accepts with some reservation
- Opposition – signaling the individual flatly refuses

In the latter case, the Project Manager will need to reassess the foundational components and associated factors and attempt again to influence the individual.

In all cases, the Project Manager will evaluate the result of the influence by reviewing, "What has contributed the most to its success?" and "What are the causes for the failure?"

For influence to be successful, the Project Manager's own behavior may undergo an evolution as personal competencies will improve by practicing various core communication skills, such as:

- Building a rapport as the basis for successful influence
- Interpreting and comprehending the significance of sensorial experience

- Being attentive in moments of silence and observing the individual's body language
- Being receptive and demonstrating this by own body language
- Seeking to determine the content of the information exchange

Effective influence is linked to the capacity to clearly state the result to be achieved.

4.10.2 Influence vs. Manipulation

Influencing is a form of persuasion and is based on mutual respect. As influence is the ability of an individual to achieve something, it can be considered a powerful tool. However, influence, according to the core foundation and associated factors described above, is not about forcing the individual to do something; it is about convincing the person that their self-esteem, interests, needs, and values are not mistreated or mishandled in any way. Influencing revolves around a sound relationship by building trust and respect between two individuals.

Manipulation is a further development along the continuum of social influence. It is a means of getting what a person wants from another individual by using tactics that may or may not be honest. There are different types of manipulation, and some tactics can be very subtle; however, they all aim to get the other person to do what the manipulator wants. These tactics put interpersonal pressure and influence on an individual for the desires of the manipulator. This takes many forms:

- Conditioning the mind
- Direct coercion and/or indirect threats
- Rejection
- Intrusive thoughts
- Subtle attacks
- Ambiguities
- Psychological warfare
- The domination game

Influence is motivated by altruism or a desire to help others, whereas manipulation is motivated by selfish desires.

4.10.3 Understanding the Power Model

As stated above,

- When Person A has the ability to influence the behavior of Person B, Person B acts in accordance with Person A's wishes
- When Person A possesses something that Person B requires, Person B's relationship with Person A is based on dependency

Social psychologists John French and Bertram Raven have described five principal bases of power:

- Legitimate – an individual's position in the formal hierarchy of an organization, with the belief that they have the formal right to make demands and to expect others to be compliant and obedient
- Reward – based on the individual's ability to distribute rewards that individuals view as valuable
- Expert – based on the individual's special skills or knowledge
- Coercive – based on fear by means of threats and intimidation, from the belief that an individual can punish others for noncompliance
- Referent – based on the individual's possession of desirable resources or personal traits

4.10.3.1 Remarks on Legitimate Power

A Project Manager, following appointment as per the project charter, will automatically have legitimate power. The scope of that power is limited to the project that others believe they have a right to control. Overuse of the legitimate power by making unilateral decisions without explanations will lead to team members experiencing an associated coercive power.

4.10.3.2 Remarks on Reward Power

Project Managers have limitations in using reward power as they rarely have complete control over salary increases, bonuses, promotions, or other organizational or financial rewards. However, praise and a mere "thank you" are powerful and extremely appreciated. The downside is when a Project Manager praises a bad performer, as this may be interpreted as a negative reward to those individuals who consider that their superior performance deserved praise.

4.10.3.3 Remarks on Expert Power

A Project Manager can be extremely effective when using expert power. By engaging with a variety of individuals the Project Manager is called upon to understand multiple project-related situations, organizational and operational issues, legal and contractual matters, and most of all, leadership and motivational theory and its application. Team members and core Stakeholders expect solutions using solid judgment. Thus, by honing the relative subject-matter expertise, the Project Manager will command trust and respect, be considered by others as a leader in those areas, and be listened to.

4.10.3.4 Remarks on Coercive Power

Project Managers must avoid coercive power as the individuals on whom it is applied will experience dissatisfaction or resentment and eventually leave. However, in rare cases, individuals need to be reprimanded for bad performance or team disruptive behavior. The Project Manager must have tangible and factual information before contemplating the use of this power.

4.10.3.5 Remarks on Referent Power

Referent power may come about without anything the Project Manager has to do. Dealing in a project with key Stakeholders and managers at different levels, individuals performing in the project may consider the Project Manager as a person to be identified with. Employing referent power alone is not an effective approach for a Project Manager who wants longevity and respect. However, when combined with expert power, it can prove to be useful.

Project Managers will engage in activities to use the Power Model effectively to achieve project goals and influence and gain the support of other people while overcoming resistance or opposition. Certain tactics can be viewed negatively when managers act in self-interested ways for their own benefit. However, is also a positive force when they allow a Project Manager to gain support for needed changes that will advance the organization.

4.11 Stakeholder Obstacles Due to Resistance

The reader will find a detailed discussion on resistance due to change in Chapter 1, "Achieving Organizational Goals", Section 1.6, "Resistance to Change".

Projects create change in the way the organization will operate, in the processes used, in the new skills to acquire, and in the size of the organization. These changes will range from a moderate shift to an upheaval in the operational environment. The Project Manager will encounter many obstacles due to organizational resistance and/or individual resistance from the stakeholder community. A lack of preparation for the change may well increase organizational and individual resistance from project Stakeholders.

4.11.1 Overcoming Organizational Resistance to Change

The Project Manager will have to face communication from managers at all levels who may resent changes to the established power relationships, line operational managers who may question the changes to be made to the organizational resource distribution, and employees who may feel uncomfortable when the change will require evolved or enhanced competencies and skills. The Project Manager must engage in strong and clear leadership and ensure that the operational entity Stakeholders are well informed and supported for the change. Establishing a clear definition of roles and responsibilities for the project and providing support for the performance of work are to be addressed.

The Project Manager must plan and conduct information meetings with stakeholder groups to establish and reiterate the organization's Objectives for the change and the strategies to deploy for a successful achievement. These meetings aim to articulate why the change is beneficial and possible and that the timeframe to bring about the change in the organization is realistic. A sense of urgency is to be conveyed by creating a convincing reason for why the change is needed and that the enhanced and changed operational processes, new skills, and new job definitions are coherent.

An organization's culture is a powerful component to its success and its declared values and is a foundation that shapes what goals to achieve and strategies to follow and provides for how things– the systems, process, and structures – get performed and determines the measures and controls and the rewards. Organizational resistance as expressed by project Stakeholders to the organizational culture's approach may be difficult or even impossible for the Project Manager to respond to and these must be escalated to the appropriate upper levels.

4.11.2 Overcoming Individual Resistance to Change

Individual resistance is principally caused by changes to be made to the way people will perform their work. Operational personnel may form the largest group affected in the stakeholder community. Project team members may also show resistance when the project constraints, new skills required, and volume of work are major impediments to the successful completion of project activities. The majority of resistance is principally caused by a shift in the individual's comfort zone, with a feeling of loss and a fear of the unknown. To overcome individual resistance, the Project Manager is to ensure that people's competencies are compatible with the performance required and that individuals have the aptitude and required experience to execute the work. Individual resistance is lowered when individuals have the perception and conviction that the performance required is aligned with reality, is challenging, and meets their values and social needs.

A sensitive area of individual resistance is one based on different or opposing cultural values. The Project Manager needs to learn and be aware of cultural and national differences in communication styles, leadership approaches, status recognition, tolerance, and many other behaviors that may not be prevalent. English native speakers use the world's business "lingua franca". This is not the case for the largest percentage of employees across the world and in international organizations. The Project Manager must therefore seek to create a shared Vision and understanding of team identity and engender an environment where people can dialogue between cultures to develop a common language and share and align work practices.

4.11.3 Converting Resistance and Confrontation into Collaboration

Resistance from Stakeholders results in the progress of the project being hampered and, in dire cases, halted or suspended.

Expectations will evolve and Stakeholders will change. Resistance will arise from different sources such as organizational, personal, financial, technical, legal, and administrative aspects. The Project Manager is to recognize the origin and record the rationale of the objections and be neutral when discussing the project and must accept that it may be necessary to modify the project Objectives and scope, while seeking a problem-solving approach and to negotiate a compromise. As a last resort, the objections and/or modifications are escalated to management and the sponsor for resolution.

Frequent meetings with Stakeholders offer crucial platforms to manage expectations and thus can alleviate resistance. Charting Stakeholders is the key to meeting at the appropriate time during the project life cycle and providing the right level of engagement and information.

Throughout the project life cycle, the project's stakeholder community changes, and consequently, the established assessment of importance, interest, and impact evolves, affecting the communication and meeting requirements. A lack of attention to this evolving situation may, in consequence, create resistance and potential confrontation.

Throughout the life cycle the Project Manager must anticipate and alleviate conditions that may lead to resistance and establish a relationship base with Stakeholders from the outset of the project so as to be able to handle confrontations of any nature as they arise. There are no easy solutions or ready-made recipes that the Project Manager can call upon to convert resistance and confrontation into collaboration as resistance will be experienced from different groups of the stakeholder community. However, there are many actions that the Project Manager can perform that can prove to be positive for creating collaboration, which would in most cases reduce or eliminate resistance and confrontation. Below is a selection of major actions that can be provided by the Project Manager:

- Planning and scheduling frequent meetings with operational departments and staff and calling on participation from all concerned Stakeholders. Mutually elaborated plans are to be developed to facilitate project Deliverables' transfer to the operational environment along with the corresponding support
- Developing with the respective performing entity a staff education plan for new skills and new operational processes
- During project execution progress, in readiness for operational handover, mutually review, assess, and consolidate with the corresponding operational Stakeholders the project plans to facilitate operational environment changes and improvements to be made. The necessary adjustments can subsequently

be made to the change plans so as to ensure coherence of approach and acceptance by the concerned Stakeholders

- Removing barriers to change, such as a lack of information, ill-defined Objectives, a lack or nonexistence of a comprehensive skill change education schedule, and other structural and organizational obstacles that people consider impediments
- Involving all Stakeholders who perform project activities to recognize that all plans are by nature imprecise and approximate, as these plans relate to work to be realized in the future. The Project Manager should encourage risk-taking and creative problem-solving
- Reinforcing the positive changes made by demonstrating the relationship between new behaviors and organizational success
- Recognizing and saluting individual changes by acknowledging new behaviors and their relationship to organizational change success
- Planning short-term "wins" and securing funding to reward (financially where possible) achievements by project Stakeholders that move the organization toward the new changed environment
- Seeking support from Stakeholders who have willingly accepted the change
- Negotiating with stubborn individuals, without discounting co-optation. Employing coercion and manipulation as a last resort

The Project Manager must then seek and establish agreement with the stakeholder individual or group by initially identifying and focusing on meeting the mutual interests of both parties and resolving ambiguities. Outcomes, results, and Deliverables to be achieved are identified and planned within the project's schedule, including the stakeholder's level of involvement and that of other Stakeholders. This will include a meeting calendar to assess changes to the agreement.

4.12 Negotiating Techniques

Negotiation is an exchange between two or more individuals or parties to reach a mutual agreement with regard to an issue of conflict.

The Sources of Conflicts and resistance in projects, as described in the sections above, are numerous, and the Project Manager must be prepared to frequently engage in reaching agreements that can be beneficial for all or some of the Stakeholders involved.

As a negotiator, the Project Manager should establish the project's requirements and needs and wants and seek to understand the ones of the stakeholder involved so as to increase the possibility of reaching a common agreement.

When negotiating in the interest of the project, the Project Manager will call upon the appropriate use of the Power Model associated with the techniques acquired for influencing (these have been described in detail in the sections above).

4.12.1 Negotiating Influence Strategies

Negotiation on project matters will require the Project Manager to be agile and flexible in thought and demeanor when addressing the issue with the respective stakeholder. Exchanges with upper management Stakeholders will require a different posture and language as compared to the one that may be employed with a team member. The major influence strategies that can be envisaged are reason and friendliness; consensus and interpersonal, and directive and hierarchical. During negotiations, the Project Manager will call upon the use of the panoply of the Power Model, conflict resolution model, and influence techniques.

4.12.1.1 Reason and Friendliness Influence Strategy

Reason relies on data and information and implicates planning, preparation, and know-how. The Project Manager is to use facts and logical arguments, as it is the strategy most often used to persuade.

Friendliness consists in showing interest, goodwill, and respect to create a favorable impression. The Project Manager will "poll" the mood of the other person before formulating a request. The strategy rests on personality, relational capacities, and sensitivity in recognizing the moods and attitudes of others.

4.12.1.2 Consensus and Interpersonal Influence Strategy

Consensus aims at mobilizing and rallying others to the Project Manager's reasoning and adds "weight" to the argumentation. This strategy establishes alliances with others and may utilize social networks to influence others.

Interpersonal influence rests on the social norms of obligation and reciprocity, and it is an art to influence others by negotiation. The Project Manager is to emphasize on the exchange of benefits or favors.

4.12.1.3 Directive and Hierarchical Influence Strategy

Directive involves the Project Manager giving orders and setting deadlines to accomplish the set desires. When necessary, the Project Manager may use visible signs of emotion.

The hierarchical strategy utilizes the formal organizational lines and the Project Manager relies on the stakeholder's direct report. In projects, this may eventually short-cut the indirect report.

4.12.2 Negotiating and Bargaining

Negotiation brings the Project Manager and the other parties together to arrive at a solution acceptable for the resolution of the conflict.

The Project Manager is to focus on the problem, not the people, and focus on interests, not demands. It is the Project Manager's responsibility to create new options for joint gain and to concentrate on what is fair.

Various alternative ways are explored, and a give-and-take decision-making process may be involved with different preferences proposed.

This may lead to entering a bargaining strategy where assertiveness and cooperation from both parties will be confronted, resulting in either a win-win or a win-lose conclusion.

A negotiation that seeks one or more settlements creates a win-win solution. The Project Manager will guide the negotiation so that both parties perceive that they might be able to achieve a creative solution to the conflict by using collaboration or compromise.

A negotiation that takes a competitive adversarial stance forces a win-lose situation, where parties see no need to interact in the future. The conflict is not resolved and the parties do not care if their interpersonal relationship is damaged by their competitive negotiation. The Project Manager must avoid being adversarial as the project's goals are of higher importance.

Bargaining may result in the best alternative to a negotiated agreement (BATNA), which would be the lowest acceptable outcome for a negotiated agreement for either or both the Project Manager and the corresponding stakeholder.

4.12.3 Decision-Making Approaches

When the Project Manager faces resistance from a group of Stakeholders, who all may have different positive or negative attitudes toward the problem at hand, it will be appropriate that decisions are made by the group. However, this may not be possible as owing to their organizational position and rank, one of the group members may decide unilaterally and conclusively on the matter.

The Project Manager should promote a more democratic approach to conflict resolution by engaging the group to vote and arrive at a decision made unanimously, following a majority or consensus. Care is always to be taken as to the repercussions when a small number of group members in a minority meet and decide for the others.

4.12.4 Approaches and Remedies for Difficult Situations

No conflict situation is easy to handle and the Project Manager must adopt a behavior that brings together all the communication skills acquired.

Great flexibility is in order to adapt a specific influence strategy to a specific person and modify the strategy depending on who needs to be influenced: upper management, sponsorship group, functional manager, peer, or team member. Care is to be given to not always use the same influence strategy with the same individual or group.

The Project Manager is to be a stalwart in recognizing and utilizing the influence possessed and not abandon easily when confronted with resistance, obstacles, and refusal. At times, the Project Manager should not shy away from utilizing influence strategies such as authoritarian, bargaining, coalition, etc.

Following any conflict resolution situation, the Project Manager is to assess the results obtained following the use of the influence strategy that was considered and its effectiveness.

As a rule, the Project Manager should not pursue an ineffective influence strategy or utilize prematurely an authoritarian influence strategy.

Success in confronting difficult situations is difficult to measure. However, the following shortlist of behaviors will go a long way in achieving success when the Project Manager engages in actions to:

- Be a good listener
- Be sensitive to the needs of others
- Be compassionate and understanding
- Emphasize harmony
- Be cooperative and not overly competitive
- Build rapport through conversations
- Advocate participation
- Communicate a desire to work together to explore a problem or seek a solution
- Seek compromise rather than dominating
- Avoid feelings or perceptions that imply the other person is wrong or needs to change
- Treat others with respect and trust

Chapter 5

Project Portfolio Management (PPM) and the Project Management Office (PMO)

Please refer to Chapter 1, "Achieving Organization Goals", for details on trans-formational and tactical Change Management; Chapter 2, "Management of Programs", for further understanding of the role of programs and projects; and Chapter 4, "Stakeholder Management and Engagement", for further appreciation of the community engaged in Project Portfolio Management and the PMO.

5.1 Chapter Overview

An organization must address its dynamic and volatile environment, be it internal or external, and manage change strategically or operationally, by responding in a proactive or a reactive manner to factors outside of its control. Change in the orga-nization must call upon the most effective way to achieve its goals. And that is by management of projects. However, because of the limits of funding, resources, and time, projects cannot be launched without assessing the organizations' ability to suc-cessfully perform them and if they fit within the wider strategic goals and Objectives of the organization.

DOI: 10.1201/9781003424567-6

The organization must constantly ask two simple questions: "Are we Doing the Right Thing?" and "Are we Doing the Thing Right?", where the "thing" in the context of this book is a project.

The focus of Enterprise Project Management is to combine the organization's Vision and Mission of the future and the operational realities of the present and map it to its pursued Objectives and structural capabilities to meet the needs of its market and customers.

Enterprise Project Management draws upon a variety of disciplines of project management with concepts, techniques, and tools to address areas such as profitability, efficiency, growth, and competitive advantage. All of these will deliver the major concerns of Stakeholders for the increase of value, maintaining and sustaining operational performance, and long-term competitiveness.

To respond to "Are we Doing the Right Thing?", an organization's challenge is to establish and maintain a structured Project Portfolio Management approach, which is key to selecting and performing those projects that will best contribute to meeting the Strategic Intent and organizational goals and Objectives.

To respond to "Are we Doing the Thing Right?", an organization must institute and operate a Project Management Office to support the project management discipline to attain a level of competence that enables the organization to define and deliver projects to the highest standards.

This chapter focuses on the above two key organizational structures. The first part of this chapter details how to establish and function in a Project Portfolio Management infrastructure, which enables and guides investing in the Right Projects, providing those projects with the right resources, and ensuring that the projects are launched and completed for the right reasons and within the right timeframe. The second part of the chapter details how an effective associated Project Management Office must be instituted so as to provide holistic support to the organization for effective and successful project performance.

5.1.1 Project Strategy and Project Portfolio Management

Business strategies essentially center on how to enable an organization to get things done in response to a changing environment and achieve results.

To the mirror questions, "Are we Doing the Right Thing?" and "Are we Doing It Right?", an organization's primary concern is to continuously examine and determine its critical issues and how its opportunities, strengths, and skills can be best employed to address these issues. A business strategy, be it at the executive level or the functional unit level, must meet/address critical issues, be aligned with the organization's Mission, and be financially viable.

Business strategies will create change, and initiatives are formulated and programs/projects are launched to make the change happen (Figure 5.1).

The strategy establishes a road map for organizational development and addresses how to develop, manage, and deliver programs/projects to reap Business Benefits

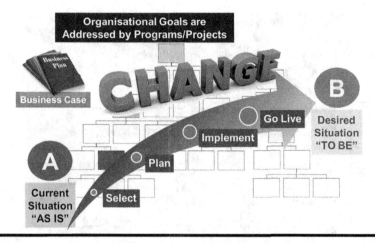

Figure 5.1   Change from As-Is to To-Be.

and value. The road map provides clarity on the organization's Objectives, creates awareness about the organization's funding/resources, and responds to a dynamic environment.

The strategic approach is to be disciplined to keep it focused and productive. It is about fundamental decisions and actions, implying that organizational decisions are to be prioritized. The business strategy clearly shifts the focus from project management to Management By Projects.

5.2 Management by Programs/Projects

Organizational Strategy drives initiatives and requires capability and capacity to achieve its goals (Figure 5.2).

Programs/projects, issued from initiatives, require constrained funding and are performed by resources that share other responsibilities alongside their project workload. To ensure coherence in order to meet the Strategic Intent(s) across the organization, programs/projects need to be assembled and assessed based on how they meet those strategic needs.

A project portfolio is a collection of programs and projects grouped together to facilitate the effective management of their performance to meet strategic organizational Objectives. Programs and projects managed within the portfolio may be directly related or mutually independent. The project portfolio represents the organization's set of active programs and projects and is an intimate reflection of the strategic goals and intents of the organization.

Project Portfolio Management (PPM) ensures that programs/projects and funding and resources are aligned with corporate strategy and Objectives. The extent of PPM is scalable and is to be tailored to the organization's environment (Figure 5.3).

Figure 5.2   From initiatives to programs and projects.

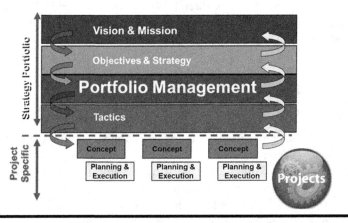

Figure 5.3   Positioning portfolio management in the strategic framework.

A key benefit of Project Portfolio Management is that it provides executives and functional managers with a synthetic view of how programs/projects contribute to the organization's Strategic Intent. PPM also assists executives and functional managers to assess the necessity of programs/projects and where funding/resources are needed.

Project Portfolio Management is a framework that translates strategy into programs/projects and aligns these to the financial and capacity management disciplines of the company.

Project Portfolio Management will provide for the periodic review, direction, and allocation of priorities and resources across the portfolio. This will consider:

- The business/functional unit or organization's strategy and Objectives
- Changes in the internal or external environment
- Business operational performance
- The status, expected benefits, and risks of all portfolio programs/projects

5.2.1 Strategic Planning and Project Portfolio Management

Strategic planning has been discussed in both Chapters 1 and 2, and the reader should refer to those chapters for further details on this theme.

Suffice to refresh in this chapter that Project Portfolio Management is directly associated with initiatives launched either at the transformational executive level and/or the Tactical Operational functional unit level (Figure 5.4).

Project Portfolio Management (PPM) is not just an administrative approach to supervising multiple programs/projects; it is a prerequisite to ensure that business value is attained and that programs/projects adhere to the strategy. It will provide management of the project portfolios and involves identifying, prioritizing, authorizing, managing, and controlling programs and projects to achieve specific strategic and operational organizational Objectives.

PPM is a process that focuses on ongoing programs/projects to validate and update the priorities and resources across the portfolio. Programs/projects are periodically reviewed to confirm their continued alignment with Objectives, and their priority is adjusted depending on their performance.

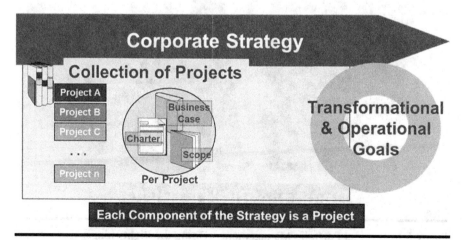

Figure 5.4   Strategy and transformational and operational goals.

A PPM system requires a robust evaluation process. Programs/projects are evaluated during their lifecycle at "stage gates". The key is the continuing relevance to the initiative's Business Case and the ability to deliver to the organizational Objectives. The evaluation process ensures that the organization stays focused on delivering the strategy and Business Benefits and that the required internal/external resources are efficiently deployed to reap the best return.

A PPM functions at a summarized level of data, gathering key information from the details of each program/project maintained at the project team level.

The major areas of concern are:

- Project performance – actual vs. planned (scope, cost, and schedule)
- Project alignment to the corporate Objectives and benefits
- Project contribution to the overall performance of the portfolio
- Impact on other projects in the portfolio

5.3 Project Portfolio Management

An organization's Strategic Intent can only be accomplished by an efficient and focused approach of Management By Projects. The Strategic Intent is translated into road maps of initiatives constituting sets of programs/projects requiring funding and allocation of resources (Figure 5.5).

Project Portfolio Management allows for an efficient allocation of funding and internal/external resources for those programs/projects currently in or to be added to the portfolio. Each program/project is subsequently assigned a contribution ranking to the portfolio's Strategic Intent based on its:

Figure 5.5   Project Portfolio Management – Doing the Right Thing.

- Alignment with the organizational Vision
- Support for the strategy and Objectives of the organization
- Reflection of the organization's cultural values
- Meaning and validity to Stakeholders

The PPM is the overarching management and governance process for the identification, evaluation and selection, prioritization, authorization, and performance review of programs/projects within the portfolio, and their alignment to the Strategic Intent and plan.

Programs/projects in the portfolio are subordinated to decisions based on their alignment with corporate strategy, viability, portfolio resource availability, priorities, and the evolution of the portfolio contents.

5.3.1 Determining the Extent of Project Portfolio Management

Project Portfolio Management requires a structure consisting of an organizational body of decision-makers and relevant information to facilitate decision-making and the exploitation of critical success factors. These have a direct and significant impact on the effectiveness, efficiency, and viability of an organization, program, or project such as:

- Effective corporate governance: decisions and management of performance against approved strategic Objectives
- Maximizing business value with clearly defined priorities and efficient use of resources
- Coherence in the implementation of programs and projects
- Significant use of competencies and skills to enhance the management of the organization

The key decision-makers, depending on the organization and whether the change is transformational or tactical, are board members, executives, relevant business managers, functional operational managers, key Stakeholders, and program/project sponsors.

The relevant information for each initiative considered to be launched as a subsequent program/project must include the corresponding Strategic Intent and plan, a validated Business Case, defined business driver measures (KPIs), total CAPEX or OPEX investment justification (ROI, NPV, etc.), cost/benefit analysis, Current Situation analysis, Organizational Impact Analysis, current and future operational plans, life cycle financial plans, as well as comprehensive high-level program/project plans.

Project Portfolio Management is a pivotal process for the successful fulfillment of Strategic Intents. Depending on each organization, the visibility given to the portfolio by all levels of executive and operational management will be a key critical success factor.

Members of the portfolio executive group or steering committee, sponsors, and key Stakeholders of all programs/projects in the portfolio are to receive pertinent and up-to-date progress/status information on the portfolio performance. Operational management needs to be informed as to the Organizational Readiness requirements they must prepare to fulfill the Business Benefits after the completion/handover of results from any subset of a program/project.

Upper executive management is to be informed of the investment/funding requirements for the medium and long term. In many cases this is done by a "rolling" quarterly financial statement covering 3–5 years or more depending on the organization.

Portfolio management is also required to be aware of evolutions in the Strategic Intent(s) and must be informed by the upper and executive management of any anticipated changes in the organization's direction.

5.3.1.1 Describing Medium- and Long-Term Vision of Project Portfolio

The principal and underlying necessity of Project Portfolio Management is the organization's objective to maximize profitability and sustainability by ensuring which programs and projects can be launched and performed to provide optimal Return on Investment.

Commercial private sector organizations, seeking to achieve tangible and financial goals, endeavor to deliver their products and services to market on time, thus resulting in increased revenue, while attaining the highest quality in their products and services, which increases market and client satisfaction.

Public sector organizations' goals are ultimately to look after the public's interests, whether in terms of improving access to their rendered services or otherwise addressing the public's concerns. These organizations will similarly strive to deliver high-quality services to their community and enhance the welfare of the nation.

All organizations will also target their internal operation and engage in cost reduction schemes through continuous improvement changes and streamlining of processes.

5.3.1.2 Defining Strategic and Tactical Programs/Projects

Programs and projects are launched to achieve the organization's strategic and tactical goals. Certain organizations will also be required to perform programs or projects in response to external clients' requirements. Thus, the aggregate number of programs and projects in realization and to be realized may well outweigh the financial and resource limitations of the organization (Figure 5.6).

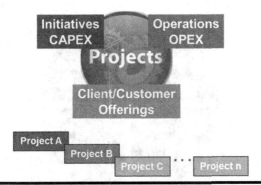

Figure 5.6   Strategic and tactical programs/projects.

It is therefore imperative that robust Project Portfolio Management be conducted within the corresponding executive and functional operational units to ensure that active programs and projects effectively achieve organizational goals.

5.3.1.3 Instituting Project Portfolio Management KPIs

The measure of performance of a project portfolio is related directly to the goals set by executives and operational functional managers. These will cover at a minimum:

- Delivery of Business Benefits
- Return on Investment
- Profitability
- Exploiting windows of opportunity
- Efficient management of operational costs

Project Portfolio Management effectiveness will be measured against the above criteria to ensure that programs/projects abide by the standard measures of success of scope, schedule, budget, and quality.

5.3.2 The Project Portfolio Management Framework

Project Portfolio Management focuses on the strategic alignment of programs/projects. If selection decisions are not made at the project portfolio level, then, by default, programs and projects are performed across the organization based on individual choices and with little regard to the impact that all programs and projects may have on the effective utilization of constrained funding and resources.

The PPM process is iterative and structured to ensure meeting Strategic Intent throughout the development life cycle of the programs/projects in the portfolio.

Programs/projects enter the portfolio following a screening and selection step and exit either at their successful completion or prematurely depending on their relative importance and rank within the portfolio (Figure 5.7).

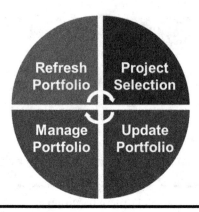

Figure 5.7   Project Portfolio Management framework.

Adequate program/project information is of greatest importance at the portfolio entry point; otherwise, it would lead to ineffective Project Portfolio Management, as it would be difficult to evaluate the benefits that would be achieved by the candidate program/project.

The key steps of the process are the acquisition of program/project information; evaluation of programs/projects; introduction to the portfolio; and monitoring of portfolio performance.

Alignment and selection can be performed efficiently by the initial acquisition of comprehensive information on programs/projects contending for a place in the portfolio.

The information gathered and collated for the project portfolio must include the business justification and value assessment and benefits targeted. Additionally, a brief description of the program/project and its scope, high-level schedule, and cost estimates must be supplied with the types and volumes of internal/external resource requirements. Risks at the program/project level must be addressed, as well as how they relate to organizational and operational risks. The time-to-market timeframe and whether the program/project is an externally enforced requirement, such as new regulatory legislation, are to be considered.

Multiple project portfolios may exist for different operational departments or business units. The choice of one portfolio versus many depends on the organization and its structure: a unique centralized project portfolio is not coherent for an organization with decentralized decision-making.

5.3.3 Project Portfolio Screening, Evaluation, and Selection

Candidate programs/projects are evaluated against predefined and agreed-upon criteria. A solid and detailed approved Business Case is imperative and a high-level

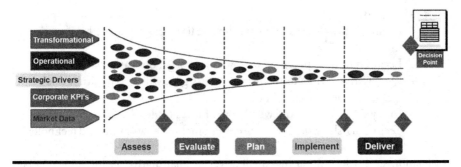

Figure 5.8   Project Portfolio Management funnel and screening

program/project scope plan are both minimum prerequisites on which decisions to continue with the candidate program/project are based.

The Project Portfolio Management process is similar to a funnel process, where candidate programs/projects proceed in phases, each phase ending as a review gate to ensure initial completeness of core information and approving continuity if the progress of the program/project remains within its delivery Objectives and maintains alignment with the Strategic Intent (Figure 5.8).

Candidate programs/projects will pass through the funnel stages and gates according to the project portfolio assessment criteria, where decisions are made to:

■ Establish a balance between transformational and operational change needs
■ Maintain consistency and alignment with the organization's KPIs
■ Balance needs to organizational capacity
■ Align funding to priorities

A first screening assessment is essential, and the following major factors are to be considered:

■ Commercial/financial factors:
  - Ability to generate future business/new markets
  - Expected Return on Investment
  - Initial cash outlay
  - Long-term market dominance
  - Payback Period
  - Potential market share

■ Internal operating impact factors:
  - Change in manufacturing or service operations resulting from the project
  - Change in the physical environment
  - Change in workforce size or composition
  - Need to develop/train employees

- Risk factors:
  - Financial risk – risks from the financial exposure caused by investing in the project
  - Legal exposure – potential for lawsuits or legal obligation
  - Quality risk – risks to the organization's goodwill or reputation due to the quality of the completed project
  - Safety risk – risks to the well-being of users or developers of the project
  - Technical risk – risks due to the development of new or untested technologies

Evaluation and selection, using qualitative or quantitative techniques, are best performed using a classification scheme. Projects can be grouped in many ways. Examples include transformational vs. operational; CAPEX vs. OPEX; organizational and geographic coverage; time to market targets; risk; skills, or technologies required; in-house vs. outsourcing; client driven vs. market driven; and so on.

Categories can be established as below, but not limited to this list:

- Increased profitability (revenue increase, generation, cost reduction, and avoidance)
- Risk reduction
- Efficiency improvement
- Regulatory/compliance
- Market share increase
- Process improvement
- Continuous improvement
- Foundational (e.g., investments that build the infrastructure to grow the organization)
- Organizational imperatives (e.g., internal toolkit, its compatibility, or upgrades)

Each project in the portfolio is quantified by its value to the business; business drivers; measurable Deliverables; and rank and priority.

A project portfolio allows for priorities to be set, investment decisions to be made, and effective resource allocation.

5.3.4 Project Prioritization/Ranking

The ranking is performed before introducing the candidate program/project into the portfolio and considers the programs/projects currently managed by the portfolio. Prioritization will rank programs/projects according to established criteria as utilized during the evaluation. The criteria can be extended to include business driver priorities; funding constraints; portfolio optimization requirements; resource capacity planning and schedule optimization; sourcing decisions; etc (Figure 5.9).

A weighting mechanism can be used to assign a score for the candidate and existing programs/projects. The ranking is then analyzed to finalize the priorities.

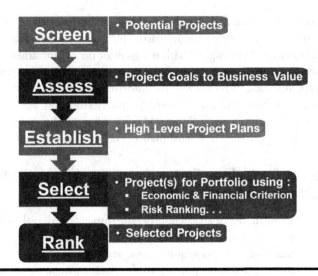

Figure 5.9 Project prioritization and ranking process.

This involves determining the portfolio mix that collectively supports the organization's Strategic Intent and achieves the defined Objectives. The project portfolio is reviewed to plan and allocate funding and resources within the overall organization's risk management philosophy.

Programs/projects that pass the initial screening are authorized funding/resources to perform additional analysis to provide the detailed data required for the next cycle of evaluation.

5.3.5 *Project Portfolio Governance Approval, Signoff, and Launches*

For their introduction into the portfolio, new programs/projects candidates that have been selected and prioritized are authorized and approved. Those candidates that have not passed the approval are postponed until the next evaluation stage gate. Ongoing programs/projects that are currently in the project portfolio may well be suspended or be ranked with a different priority. Care must be taken to factor in the human and psychological impacts caused by these changes to ongoing programs/projects, especially those that need to be terminated.

Following the selection and ranking, candidate and ongoing programs/projects are classified as "pursue", "go", "no-go", "terminate", and "postpone". New "go" entrants to the portfolio according to their ranking position are authorized funding by an executive group (steering committee or another group) and a corresponding project charter is established. A project core team and core key resources are assembled to develop a detailed project plan to be integrated into the portfolio delivery plan.

The project portfolio is then synchronized with all approved program/project plans to constitute a project portfolio master plan and schedule illustrating the financial and resource plans, aligned with the goals of the organization.

The project portfolio should also include those programs/projects that are in the evaluation step, as these candidates require funds/resources to further develop their conceptual plans, such as a Business Case or feasibility study.

Program/project sponsors and Project Managers are subsequently informed of the decisions taken by the portfolio governance, and work authorization is issued.

5.3.6 Project Portfolio – Status, Progress, Adjustment, and Maintenance

The nature of transformational strategic programs/projects is that they evolve, and new subset projects are created. The project portfolio is consequently adjusted for these new entrants. These would require to be screened before launch, although they would be assigned a similar priority to the parent program. However, other programs/projects will need to be reassessed to accommodate the new entrants.

Ongoing programs/projects are monitored for their performance at key portfolio review gates. Adjustments are made following a similar criterion to that used for selection. Individual programs/project progresses/statuses/forecasts are assessed to ensure that the distribution of funds/resources is adequate and contributes to the overall Strategic Intents. Large, long-duration, and/or complex programs/projects may be further broken down to generate earlier operational measurable benefits.

In the cases when a program/project is "terminated" or "postponed", the associated costs are to be considered. A balance is to be struck between the incurred costs and the forecast funding required "to complete" with costs such as the "sunk costs", staff reassignment costs, demobilization costs, contract termination costs, etc. Special care should be given to the impacts on the psyche and the morale of all those concerned with a "termination" or "postponement" decision.

Project portfolio evaluation and adjustment can be performed outside the established review gates. Programs/projects in the portfolio may have to be re-evaluated as the result of major budgeting cycle decisions, following major changes in scope, cost, or schedule, the occurrence of risk events part of the contingency plan, or any other incident that modifies seriously the current plans. New candidate programs/projects proposed between review gates need also to be evaluated for introduction to the portfolio.

5.3.7 Portfolio Manager Roles and Responsibilities

The Portfolio Manager's knowledge and actions revolve around key major areas:

- Benefits Realization – ensuring that the ongoing programs/projects within the confines of the project portfolio are all aligned with the organization's goals and will achieve the defined benefits

- General management skills – in addition to managing the immediate project portfolio team members, having the business acumen and ability to engage with program/Project Managers, executives, and functional operational managers
- Product and process development and continuous improvement – possessing a solid understanding of key scope contents of change programs/projects present in the project portfolio
- Program and project management methods and techniques – competent and skilled in the discipline of project management

The Portfolio Manager must possess an advanced understanding of project and Program Management and a thorough understanding of high-level project management reporting to determine if management approaches are lacking or failing.

The Portfolio Manager, typically a senior manager or senior management team, is responsible for monitoring the status of ongoing projects. Monitoring consists of:

- Playing a key role in project prioritization and ensuring that there is a balance of components and that they align with the strategic goals
- Providing key Stakeholders with a timely assessment of portfolio and component performance, as well as early identification of (and intervention into) portfolio-level issues impacting performance
- Measuring the value to the organization through investment instruments, such as Return on Investment (ROI), Net Present Value (NPV), Payback Period (PP), etc.

5.3.8 Defining the PMO Organization to Support PPM

Please see Section 5.7, "The Project Management Office", for a detailed exposé of the PMO.

While Project Portfolio Management responds to "Are we Doing the Right Thing", which ensures that the appropriate programs/projects are performed according to defined goals and Objectives, the Project Management Office focuses on "Are we Doing the Thing Right", which ensures that the discipline of project management is of the highest efficacy.

To fulfill its supportive role in the project portfolio, the PMO must participate fully to:

- Establish an organizational focus on improvement in project management competency
- Provide organizational support to achieve Strategic Intents
- Develop and/or enhance project management skills and knowledge through training, coaching, and mentoring

- Ensure consistency and uniformity in project development
- Make available project tools and techniques for project planning and realization
- Reduce project funding overruns and enhance project schedule delivery
- Increase customer satisfaction through the achievement of Business Benefits
- Provide centralization for the project management practice and discipline

5.4 Governance in Project Portfolio Management

Project Portfolio Governance is established by a governing body to make decisions about investments and priorities for the portfolio and ensures that the portfolio management processes are followed to sustain the organization. Governance processes must use a framework to align, organize, and execute the activities of programs/projects in a collectively coherent manner to achieve the defined organizational goals and fulfill the Strategic Intent.

Project Portfolio Governance will provide:

- Authority: a single body decides
- Guidelines: for project selection and funding
- Goals: clarity for strategic and operational changes
- Discipline: setting and adhering to decisions
- Data: collecting and disseminating accurate information

The Portfolio Manager will provide the appropriate governance processes to support the achievement of the organization's strategy, ensuring projects are aligned with evolving transformational and tactical Objectives.

A governance group is comprised of individuals drawn from the organization. These individuals originate from upper/executive management, functional operational management, and program/project management. Key Stakeholders and external consultants may also be part of the governance group when it is deemed necessary.

Members of the governance group must provide their commitment and availability to fully fulfill their roles in the alignment, evaluation, introduction, and monitoring of the project portfolio. Key tasks to ensure the effective management of the portfolio include but are not limited to:

- Ensuring that the project portfolio is aligned with the Strategic Intent and business Objectives
- Selecting programs and projects that will deliver the required results
- Efficiently managing the project portfolio
- Planning the portfolio program/project workload on an ongoing basis
- Determining the impact of the project portfolio on the corporation's resource capacity and ongoing operations

- Negotiating resource and capacity issues and resolving problems at the organizational level
- Selecting competent and accountable sponsors who commit sufficient time to their projects and focus on their Stakeholders' interests

The governance group should also ensure that projects have:

- Clear critical success criteria
- Key performance indicators
- Clear mandates, roles, and responsibilities for Project Managers
- Appropriate and available project management processes and tools
- Change, risk, and issue management policies and standards
- Timely and reliable project reporting
- Escalation criteria for significant issues

5.4.1 Corporate Strategy and Governance – Project Portfolio Stakeholders

The project Portfolio Manager is responsible for the portfolio management process and will engage with key Stakeholders.

- Portfolio review board: will dictate the framework, rules, and procedures for decisions
- Project Managers: to provide single-point accountability for projects in the portfolio
- Program/Project Management Office (PMO): to provide coordination and management of the portfolio components
- End users: internal and external Stakeholders who will benefit from successful program/project implementation
- Executive managers: who convey the strategic goals to portfolio management
- Operations management: to balance the need for ongoing operational performance with resource provision to portfolio projects components

5.4.2 Project Portfolio Governance Approval, Signoff, and Launches

The governing body will give final approval on program/project selection and conduct a management sign-off with the pertinent Stakeholders and subsequently issue the program/project charter. Each approved program/project is ascribed a rank, and key resources are assigned. Program/project initial funding is released, and management reserve and variance ranges are determined.

The key project drivers are documented, including goals to be achieved, decision points, assumptions, and risks.

5.5 Project Portfolio Consolidation of Project Data

Project Portfolio Management will collect program/project data to conduct key activities of consolidation, capacity planning, schedule optimization, and aggregate funding requirements. The data collected at the outset from an individual program/project will detail the planned schedule, resource loading, and associated costs. Capacity planning is a major concern to be addressed, as well as the evaluation of the resource capacity needed by the organization to meet the demands of the programs/projects in the project portfolio.

Program/project initial plans and progress and status data are compiled, and total portfolio consolidation is performed. This consists of but is not limited to:

- Aggregation of project plans to:
  - Establish project portfolio master plan schedule and timeframe
  - Develop cumulative resource charts
  - Determine collective program/project costs
- Rearrangement of program/project priorities to:
  - Review program/project progress and status against defined goals and assess the evolution of business changes
  - Establish modified program/project schedules and portfolio master plan
- Reapprove project portfolio contents to:
  - Reassess funding distribution
  - Reallocate resources where necessary
  - Secure management sign-off on changed portfolio master plan

Project Portfolio Management consolidation ongoing activities will entail:

- Evaluating and determining the performance of portfolio components and alignment with Benefits Realization (financial and nonfinancial)
- Determining resource capacity availability and constraints for portfolio components
- Performing portfolio components risk analysis according to the organization's risk criteria
- Reviewing portfolio components ranking and determining which should receive the changes in priority within the portfolio and reassigning these according to their status
- Identifying portfolio components to be suspended or terminated based on their status and the result of rebalancing the portfolio master plan

5.5.1 Project Portfolio – Capacity Planning and Management

Capacity is the maximum amount of work that the organization can complete in a given timeframe with the resources and funding availability and constraints.

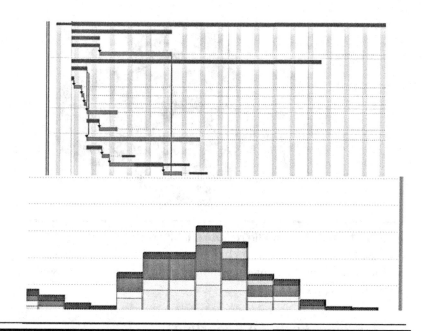

Figure 5.10   Project portfolio – capacity planning and management.

The Portfolio Manager, together with other organizational entities, must develop a sound, structured, and well-thought-out portfolio management capacity planning process that best suits the organization and coordinates well with other processes (Figure 5.10).

This demands that at each organizational entity, managers and supervisors maintain a rolling 12-month capacity plan of all the human, material, and equipment resources within their area of responsibility. All resources must be coded with unique profile references to allow for consolidation and grouping of the same resource capability. For example, a human resource with a profile of "mechanical engineer with 5 years minimum experience in steel structures" will be coded HR0125, and all the human resources across the organization with the same profile will also be coded HR0125. Similarly, for the management of equipment, a profile for a "six-wheeled 5-ton loading truck" will be coded EQ0711, and all other equipment resources with the same profile will also be coded EQ0711. Without what may seem like a very heavy profile coding structure, it is impossible to aggregate the organization's total resource capacity and availability.

The profile coding structure must also be used for resource identification in all program/project plans in the project portfolio and be utilized in the project management software. Otherwise, it is impossible to aggregate the cumulative resource needs for the project portfolio and conduct any consolidation and take decisions on resources and funding.

Figure 5.11   Project portfolio reporting framework.

The author is open to any suggestion to address capacity and resource planning using another approach.

5.5.2 Project Portfolio Reporting

Project Portfolio Management ensures that the portfolio stays aligned with business Objectives. This involves following a continuous process by which programs/projects are evaluated, prioritized, selected, and managed at formal points such as portfolio gate reviews. As noted above, this is also performed outside the formal points (Figure 5.11).

Project portfolio review gates are to be established according to a convenient frequency, depending on the organization. Monthly and quarterly reviews are common for transformational change programs/projects and usually are aligned with other corporate reviews based on either the civil or the financial calendar year. However, weekly reviews are necessary for operational change programs/projects as these are often for short durations of weeks and months.

Programs/projects provide the key performance indicators, progress, and forecast prior to the review gate. Portfolio management ensures alignment with the business strategy and effective resource utilization. Special attention is given to Benefits Realization (after handover/deployment).

The Portfolio Manager conducts the gate reviews, first at the program/project level to:

- Assess performance vs. Business Goals
- Appraise resource allocation: human, materials, and equipment
- Ratify project status in the portfolio
- Review project priority ranking
- Sanction decision to continue

After the consolidation of all program/project status data, the project portfolio is reviewed to:

- Determine overall portfolio progress
- Assess anticipated resource needs
- Realign projects to business evolution
- Establish decision criteria for the next gate(s)
- Early termination decisions

Portfolio reports may include, but are not limited to the following:

- Performance reports, including dashboards
- Resource capacity and capability reports
- Accrued costs incurred
- Variance reports and forecasts on resource and funding needs
- Portfolio risks and issues originating from programs/projects
- Portfolio component recommendations
- Governance recommendations and decisions
- Executive summary to concerned Stakeholders

5.5.3 Portfolio Management – Manage Portfolio and Maintaining Alignment

Decisions are made throughout the Project Portfolio Management process to optimize the overall contribution of the components to the organization. The Portfolio Manager must have a clear understanding of the organization's Vision, Mission, and strategy, be it transformational or tactical, to aid in the optimization of the project portfolio and understand how to relate the strategic goals and Objectives with the portfolio component plans to achieve the defined goals.

Portfolio status is assessed, and the initial focus is on the impact of business forecasts on the Strategic Intent. Portfolio master schedule funding/resource utilization and its effect on portfolio performance are analyzed. An evaluation is performed seeking out-of-range variances in scope, financial and schedule performance, and risks, and comparing these to the benefits and organizational value (Figure 5.12).

Program/project dependencies are reviewed and priorities are adapted to the previously set Project Portfolio Management criteria. This will also involve classifying certain programs/projects as "terminated" or "postponed".

Portfolio management is to be enabled to make recommendations to ongoing programs/projects for more efficient use of funding/resources in the portfolio and request an adapted schedule and cost plans to accommodate these recommendations.

Arbitration will often be required to resolve funding/resource conflicts between programs/projects. Thus, an up-to-date portfolio master schedule and resource plan

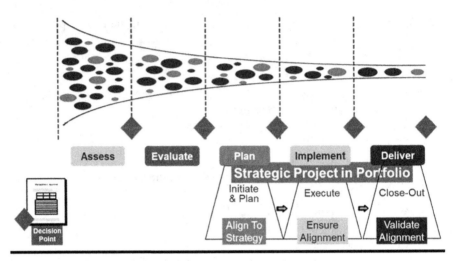

Figure 5.12   Project Portfolio Management stage gates.

are imperative. Program/Project Managers need to fully understand the ongoing evaluation/adjustment criteria to contribute efficiently to generate key information for portfolio management decisions.

5.5.4 *Project Portfolio Management – Refresh, Assessment, and Analysis*

Senior executives (or a steering committee) participate in the management of the project portfolio and are present at the gate reviews. Key to the success of portfolio management is their "big" picture view of the Strategic Intent(s). They sanction and communicate approvals and priorities and should be conversant with the future costs necessary to obtain organizational benefits, such as "cost to complete" and operations and maintenance costs.

The project portfolio is refreshed by retiring completed programs/projects and suspending or terminating any portfolio component. Pending programs/projects may then be assessed for their entry into the project portfolio, and resource and funding needs are then recalculated.

The portfolio management team is to disseminate key program project information across the enterprise. This follows the review gates but should also be performed when key Deliverables are completed or close to completion and are at the stage for operational handover. Organizational Readiness is a major concern in delivering value to the organization.

The Project Portfolio Management process must include a mechanism to capture, collate, and consolidate the lessons learned on the performance of the portfolio management and propose ways to continually refine the process.

5.6 Project Portfolio Management – Effective Performance

The justification of a Project Portfolio Management process can only be measured by its contribution to achieving Business Goals. Operational Effectiveness and achieving Business Benefits are the ultimate goals of Project Portfolio Management.

Project Portfolio Management constantly evaluates the portfolio components against the defined expectations to ascertain if:

- Strategic Intents are unchanged
- Projects are in alignment
- Project budgets are within the range
- Operational Benefits are within limits
- Project schedules are within range
- Technical issues are addressed
- Project risks are managed suitably
- Organizational risks are assessed
- Internal/external constraints and interfaces are managed appropriately

Operational performance is to be monitored and evaluated to ensure that the forecasted benefits are realized. Operational managers are to provide the project Portfolio Manager with feedback on actual performance and if and how actual benefits are in fact reached. Hard financial benefits, such as cost reduction programs, are not too difficult to measure, whereas soft financial benefits, such as a skills/competence increase program, will pose difficulty.

Of all the attributes of a successful portfolio management process, the human element is the highest on the list. The portfolio management group must assess its own "soft" skills, as its members have to interact with a variety of individuals and where diplomacy, psychology, sociology, and other human sciences are constantly used. The portfolio management team members need also to assess their business, analytical, and decision-making skills and seek improvements where necessary.

5.6.1 Project Portfolio Reporting Evaluation Post Deployment of Projects

Effective Project Portfolio Management focuses on ensuring that the transformational or tactical change has been deployed successfully in the organization's operational environment. To determine if Business Benefits can be attained, a review and an evaluation of completed programs/projects will ascertain if the component Objectives and Deliverables have been achieved and whether program/project scope contents have changed.

Furthermore, the receiving functional entity evaluates how the Deliverables have enabled their operational performance.

The benefits of Project Portfolio Management make business sense:

- Efficient use of corporate funding/resources and the management of performance against Strategic Intents and Objectives
- Improved business performance through clearly defined priorities
- Eradicating waste from non-strategic efforts which draw upon scarce funding/resources
- Greatly enhancing success in the performance of programs/projects
- Maintaining and sustaining highly motivated employees who share the company's achievements

5.7 The Project Management Office

The Project Management Office responds to the question, "Are we Doing the Thing Right?". As highlighted several times in this book, transformational and tactical changes are accomplished by programs and projects that are launched to achieve organizational goals and create value for the organization through the realization of Business Benefits.

Senior executives, overseeing CAPEX and transformational changes, must deliver organization-wide results, generate growth, and pursue opportunities aligned to the Strategic Intents over medium- and long-term timeframes.

Functional managers, managing the fiscal year OPEX and tactical changes, are continuously evaluated for their ability to maintain and sustain operational performance and focus on continuous improvements to ensure the highest quality of the organization's products and services.

It is self-evident from the above that no changes in the organization can be realized effectively if projects are not successful in delivering their results within the target timeframe and the allocated funding.

There is a multitude of reasons as to why projects fail to deliver, just to name a few:

- Incomplete or weak Business Case
- Deficient scope definition
- Inadequate scope content management and uncontrolled scope content evolution
- Changing priorities
- A lack of or no management buy-in
- Weak project team management
- A lack of or nonexistent project standards, methodologies, and procedures
- Poor or nonexistent documentation
- Absence or misuse of standard project management tools and techniques
- Sporadic and/or inaccurate project reporting

- Insufficient resource demand visibility
- A lack of risk management rigor
- Insufficient project procurement visibility

All the abovementioned, and other, factors will hinder the organization's ability to achieve project schedule, budget, and quality goals, as the project management discipline is inadequate for Doing the Thing Right.

To raise the level of the project management discipline, improve the project delivery in scope content, budget, and schedule, provide project information to executives and functional managers, and institute a project decision process – a structure such as a Project Management Office (PMO) is of great necessity and will accompany the formal Project Portfolio Management as described above.

5.7.1 Project Office vs. Project Management Office

A Project Office is an organizational unit in which the Project Manager leads the project team in planning and conducting the project. The Project Office provides administrative support, subject-matter expertise, methodologies, tools, and techniques, and manages the repository of all project documents. The Project Office can be a simple one-person office, a complex suite of offices, or a virtual location. Irrespective of the scope, the Project Office concentrates on one project only. Every project ever performed has always been managed by a Project Office.

A program/Project Management Office is instituted when two or more projects are to be administered from a single location and/or by a single responsible manager. The PMO will provide administrative support, subject-matter expertise, methodologies, tools and techniques, and tools to be used by projects to manage the repository of all project documents. The physical location of the PMO in the organization may vary from enterprise to enterprise.

There are many types of PMO that can be implemented, and this will be reviewed in detail in the sections below. There may be many PMOs in a given organization, from an overarching central executive–level PMO that addresses transformational change programs to individual PMOs at the functional operational level to address specific tactical change programs and projects.

Suffice it to state that a PMO, irrespective of its size or location, will provide a large range of services to the project management community and to the organization's management at all levels, in the latter's pursuit to achieve its change Objectives by deploying high quality and successful management of projects.

5.7.2 Purpose and Goals of the Project Management Office

The mandate and purpose of a PMO as an organizational entity is to centralize and coordinate the management of the project disciplines under its domain. A PMO's purpose can be extended to oversee the management of projects, programs, or a combination of both; care must be taken in the definition of the verb "to oversee", as it has a wide range of definitions such as *to supervise, run, control, manage, direct, handle, conduct, look after, be responsible for, administer, inspect, preside over, keep an eye on, and have or be in charge of.* As is further developed in the following sections, clarity in defining the scope of the PMO is fundamental and must be aligned with both the Maturity Level of the organization and the type of PMO to operate.

There are many compelling reasons to operate a PMO. In conjunction with Project Portfolio Management, the PMO provides an overview of all program/project delivery activity within the organization. By deploying standard project management methodologies, the PMO assures better continuity and maintenance of these standards for all programs and projects, which will drive improvements in schedule and budget predictability, quality project Deliverables, and employee morale. The organization will subsequently experience increased Return on Investment and greater satisfaction from executive management and internal functional units, while overall management of project "productivity" will increase in conjunction with professional skills development for the project management community.

The major goal of a PMO is the establishment and enhancement of project management professionalism and requires senior management support. (*The discussion of a PMO sponsor can be found below in this section.*)

This is achieved by instituting and deploying consistent project management methods, systems, processes, tools, techniques, and metrics, and providing centralized project management expertise. Achieving these goals will equip the organization and the programs/projects with a structure and a professional environment where:

- Repeatable project delivery process can exist
- Enhanced estimates can be made based on standards and historical data
- Coherent management of priorities is adopted
- Improved resource management across projects can be achieved
- Reduction of project waste can be accomplished

5.7.2.1 Project and Enterprise-Focused Functions of the PMO

The PMO provides services to a wide audience and will interface directly or indirectly with executive management by enhancing the achievement of Business Goals and Strategic Intents. The PMO will provide operational management with a coherent methodology to integrate functional and project activities and enable an increased probability of meeting their needs and expectations.

The key enterprise focus will be on the projects' core community, providing Project Managers and the project team a variety of support services to assist in the management of projects, operations and administrative support, and training and career development in the discipline.

5.7.2.2 Benefits of Establishing a PMO

The benefits derived from instituting and operating a PMO are for the most part seemingly intangible. However, by establishing an enterprise focus on improvement in project management competency, benefits can be measured by recording the progressive enhancements in project schedule delivery and quality, the reduction of project funding overruns, and the increase in functional unit satisfaction by the achievement of their Business Goals.

In providing organizational support to achieve Strategic Intents, benefits from a PMO will also be attained as it deploys the development and/or enhancement of project management skills and knowledge through training, coaching, and mentoring.

Most importantly, the major benefit of a PMO is its ability to provide centralization for the project management practice and ensure consistency and uniformity in project development.

5.7.2.3 Different Levels of PMO in the Organization

There is no single structure of a PMO; there is a step-by-step evolution that corresponds to the Maturity Level of the organization. The structure is mapped to the Maturity Level of the organization and provides a choice on how to progress to higher levels. The Maturity Levels are different from those identified by the PMI.

The PMO structure selected must satisfy the organization's needs, and in all cases, the PMO must have a sound business and organizational purpose.

5.7.2.4 Maturity Level of the Organization

The Maturity Level of the organization is evaluated by employing one of the many different industry-recognized questionnaires. The most known ones are the Project Management Institute's (PMI) OPM3 and the Office of Government Commerce's (OGC) P3M3. Both can be purchased.

PMI's OPM3 (Organizational Project Management Maturity Model) is a means to understanding and assessing the ability of an organization to implement its high-level strategic planning by managing its portfolio or portfolios and then delivering at the tactical level by successfully, consistently, and predictably managing programs and individual projects. It provides a method for organizations to understand their project management processes and measure their capabilities in preparation for improvement. OPM3 results can then assist organizational leaders in developing a road map that the enterprise can follow to improve performance.

OGC's P3M3 (Portfolio, Program, and Project Management Maturity Model) is owned by Axelos, a joint venture between the UK Government and

Capita. The assessment establishes five levels of maturity, in response to the questionnaire results.

- Level 1: Initial – the starting point for the use of a new process
- Level 2: Repeatable – the process is used repeatedly
- Level 3: Defined – the process is defined/confirmed and institutionalized as a standard business process
- Level 4: Managed – process management and measurement take place
- Level 5: Optimizing – deliberate process optimization/improvement

P3M3 is made up of three models, which can be assessed individually or as a group: Portfolio Management, Program Management and Project Management (Figure 5.13).

For each of the models, P3M3 reviews the maturity and performance against seven perspectives:

- Management controls
- Benefits and/or requirements management
- Financial controls
- Stakeholder Management
- Risk Management
- Organizational governance
- Resource management

The results of the P3M3 assessment allow the organization to decide on its current status and establish a plan for instituting, deploying, and operating a PMO and its

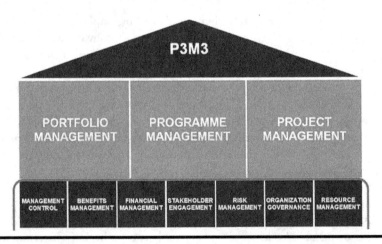

Figure 5.13   OGC's P3M3 – portfolio, program, and project management maturity.

evolution and progression. This is done by first conducting a Current Situation analysis of "As-Is", which produces an assessment of the current state of the organization and which areas will be impacted. This is followed by a description of the Vision of what to achieve, the desired situation of "To-Be". A GAP analysis and transition plan will then be established to define the products and services, processes, skills and competencies, and infrastructure to be reviewed/enhanced/introduced. Goal setting and measurement tools to use are then put in place.

5.7.3 Key Factors for PMO Success

Introducing a PMO in the organization is a forthright way to demonstrate that project management professionalism is rated highly in the enterprise. However, creating a PMO and not providing the support and structure with essential and significant means for it to operate is not only a waste of time and money but also creates negativity and rejection by management as it does not consider that the PMO has any value.

Before the launch of a PMO, there must exist executive management buy-in and accompanied support and an assigned PMO sponsor (see below). The need for the institution of a PMO must be substantiated and quantified by a convincing analysis of the organizational needs and a supportive Business Case that will set management of expectations as to the goals that the PMO will target to achieve.

The PMO, once approved to proceed, must be initially staffed with a PMO manager and at least one subject-matter expert in the field of project management and associated methodologies. Dedicated facilities, equipment, and infrastructure must also be made available.

The deployment of project management within the organization and the accompanying PMO may not be widely perceived as positive by everyone. This is true for any cultural and organizational change initiative.

Thus, the PMO must be announced to the organization in a formal manner, preferably by senior management, to ensure and strengthen its stature as an important support group to the management of projects to achieve transformational and tactical changes.

Without a solid foundation for the PMO, there will be key barriers to its success stemming from:

- No executive management buy-in
- Politics and power struggles hindering PMO functions
- Unclear or misunderstood purpose of the value of the PMO
- Unrealistic expectations of Objectives and goals to be achieved by the PMO
- Consideration of the PMO as an overhead and of little value
- Perceptions of the PMO as being too authoritative by the project management community

5.7.3.1 Assignment of the PMO Sponsor

The PMO must report directly to a sponsor if greater success is to be achieved. No matter what type of PMO is established, there is a greater chance of success if the PMO is sponsored as high as possible within the organization. High-level executives are ideal, such as the CEO, CIO, CFO, or the like, when the PMO is at a strategic level for transformational change programs, as well as senior functional managers who have the responsibility for operational performance and managing tactical changes for sustainability and/or continuous improvements.

Depending on the nature of the organization, be it private or public, and the industry sector in which it operates, much of the organization may consider the PMO as a luxury and an overhead. If the enterprise is in an industry where most of its functions are purely operational, such as industrial production, mining, utilities, or manufacturing, its continuous focus is on cost efficiency and product quality, and continuous improvements are a mainstay of maintaining and sustaining its operational performance. Business unit managers and functional heads may not consider the formal management of projects by a PMO structure to have value. Whereas, if the enterprise's functions revolve around the development and creation of infrastructures, products, and services, such as civil engineering, construction, and other solution-selling structures, then management of projects is of fundamental importance and a PMO would be considered to have value.

Thus, until the PMO can establish the value it is delivering, it is basically as strong and as protected as the person or persons that are sponsoring it.

The organizational power of the PMO, derived from the sponsor's organizational position and rank, may evolve when there is a sponsorship change. Therefore, the PMO must produce positive results within a short timeframe and maintain quality support services to the project management community, and quantify its value so that the new sponsorship can continue its support as provided by the previous sponsors.

5.7.3.2 Identification and Involvement of Influential Stakeholders

The PMO manager must strive to maintain constant communication with the community it serves. Beyond the obvious direct communication with the PMO sponsorship, direct lines of communication must be fluid between the program/Project Manager(s) and project team members who perform on the project and who apply the standard project management practices and procedures provided by the PMO. The individual project sponsors who have authority over project funding and resource allocation are to be included in a larger management of the project community.

At a wider level of communication, executives and senior management steering committee members and strategic decision-makers are to be provided with the progress and positive results of the PMO. Functional and resource managers who provide specialized skills and other resources to projects must be equally included in the communication loop to ensure that they comprehend their participation in the execution of programs/projects. And the customers/clients/end users who establish program/project needs and requirements and take delivery of the results for operational performance needs are to be ignored.

Sustaining stakeholder buy-in for the PMO is a challenge, as instituting a PMO requires different behaviors of those who perform on projects and those who are the clients of the project. Stakeholders must be convinced as to why a Project Manager requires time to create a project definition, perform risk management, apply rigorous scope Change Management, etc.

The PMO manager must hold frequent meetings and workshops as needed with Stakeholders across the organization to convey its functions, methods, and delivery successes to gain and maintain their support and buy-in.

5.7.4 Project Management Office – Role, Type, Functions, and Organization

The principal role of the PMO is to ensure that the project management community can perform to the highest quality that the discipline requires for success. This revolves around three pillars:

- Efficient and functional methodologies, standards, and processes
- Subject-matter expertise for project support
- Project management community competence and skill development

For the first pillar, the primary role is to develop, deploy, make available, and maintain the management of project methodologies, standards, and processes, ensure their usage across all projects, and monitor their effectiveness. Continuous improvement is to be applied to the body of methodologies, by collecting feedback, comments, and suggestions from the Project Managers and teams. A central library for these methodologies and standards (including templates, forms, and checklists) is to be provided, and expert support is made available for their learning and deployment. As the PMO functional level matures, a project knowledge management system should be made available to capture and disseminate project lessons learned and other helpful information for future projects.

The second pillar will provide subject-matter expertise for the use of specific project tools and techniques. This will alleviate the workload on Project Managers and core team members, as the work to be performed by the subject-matter expert is usually either of a short duration or spaced throughout the life of the project. The aim of the subject-matter expert would be to:

- Provide estimating and budgeting
- Conduct the development of plans and schedules

- Assist in project risk management
- Provide status updates and back-office administrative support
- Perform variance analysis
- Maintain a project repository and store status/progress/forecast reports
- Establish and maintain project-related software tools, such as project management software and supporting software, as well as industry standards

The third pillar concerns directly the project management community, with Project Managers first in line. Project team members are also involved, and the organization, in conjunction with the PMO, can decide on extending competence and skill development to non-project individuals and functional units. For Project Managers, the PMO is active in the development of professional Project Managers. Maintaining a database of able and available Project Managers and documenting their skill sets and experience are central to this pillar, as it allows the PMO to focus on targeting the required training/professional development. This development is multifold and can cover:

- Identification of the competencies needed by high-performance Project Managers
- Provision of training and development plans to Project Managers and team members
- Establishment of project leadership and team management training and development
- Support of career paths
- Promoting project management certifications
- Provision of expert assistance in the form of coaching for functional staff involved in projects

The type of PMO must correspond to the present project management maturity of the organization. Its institution and how it will operate will depend on key factors such as:

- Business needs
- Vision and goals of the sponsor
- Political environment
- PM Maturity Level
- Organization size
- Number of projects

The PMO structure and span of responsibility will evolve from a simple Project Office to an organization-wide Centre of Project Excellence.

On first introducing the PMO, the organization must have a conservative approach, as declaring a type higher than the one compatible with its Maturity Level is unfortunately a path to failure. This book considers four types of PMO levels (other publications have a different number; however, all cover the same principles). The types are:

- Basic PMO
- Standard PMO
- Advanced PMO
- Centre of Project Excellence

5.7.4.1 Different Levels of PMO in the Organization

I. A basic PMO fulfills a project management role in the organization. The basic PMO typically oversees projects that require diverse schedules to be consolidated into a master program schedule. The choice of a basic PMO is driven by the result of the maturity assessment identifying that:
   - Previous project management practices have been sparse
   - A need exists to develop and maintain a common set of processes and methodologies
   - Meager "services" are provided to projects across the organization

   The key function of the basic PMO is to collate/integrate and disseminate project management methodologies and processes. Program Managers may or may not report directly to the PMO, while project team members tend to be assigned by external functional managers.

II. A standard PMO articulates a "management of projects" approach in the organization. It is a progression from the basic PMO. The standard PMO oversees multiple projects within a business unit across functional departments. The choice of a standard PMO is driven by an increase in the Maturity Level of project management. It is now recognized to be more effective as it provides consistent and comprehensive project support to assist Project Managers, is better structured to project management competence development and career pathing, and importantly, the organization begins to acknowledge the value of the PMO's services as the PMO's visibility is enhanced. Project Managers may be assigned to the PMO and they may or may not report directly to the program or PMO manager. Project team members are more aligned with the PMO mode of operations, while PMO support staff are added progressively according to requirements and necessities.

III. An advanced PMO instills a "Management by Projects" culture in the organization. This allows the organizational entity, be it at the executive or the business unit level, to manage its entire collection of programs/projects as project portfolios and provides management with a view of the organization's program/project activity from a central source. To reach this level, the advanced PMO has been firmly recognized by management as a strong organization for the delivery of value for projects, has the ability and capability to sustain consistency in project management methodologies, participates effectively in fulfilling the company's Strategic Intents, and is a strong base for continuous improvement and growth.

Operating as an advanced PMO, Project Managers will usually report directly to a program or PMO director, and project team members are proficient in the use of project management methodologies as deployed by the PMO. PMO staffing now consists of positions by subject-matter specialty, and competency and skill development training and education are firmly established in the organization along with certification programs.

IV. The Centre of Project Excellence (COPE) is the "best in class" for organizations that aspire for total Management by Projects to ensure the achievement of Business Benefits in the pursuit of Strategic Intents. The COPE institutes top professionalism for the delivery of programs and projects across the organization. The COPE director reports to a vice president or executive team, or higher, and will promote excellence throughout the organization in:

■ The project management discipline
■ Skills and competencies in project management
■ Project management methodology
■ Standards
■ Policies and procedures
■ Process control, support, and improvement

5.7.4.2 The PMO Core Organizational Structure: Roles and Responsibilities

The core PMO organizational structure and functions are the same for whatever type the enterprise decides to operate. The PMO manager will need to build a core team to fulfill the functional areas of the entity. These functional areas are:

■ Development and dissemination of processes, standards, and methodologies
■ Providing project support expertise
■ Project management competency and skill development

 I. Development and dissemination of processes, standards, and methodologies
The primary role is to
  – Develop, deploy, make available, and maintain the management of projects processes, methodologies, tools, and techniques
  – Provide a central library for these standards (including templates, forms, and checklists), and expert support for their deployment
  – Incorporate lessons learned on projects into the project knowledge management system and methodology
 II. Providing project support expertise
  – Assist in the development of project plans and schedules, assist the development of plans and schedules

- Provide back-office administrative support and maintain a project repository of project progress and forecasts, and when requested, provide status updates and perform variance analysis
- Establish and maintain project-related software tools and conduct the acquisition of project management software and supporting software, as well as industry-related standards

III. Project management competency and skill development

The PMO is to be active in the development of professional Project Managers and is to maintain a database of Project Managers' skill sets and experience.

The focus is to be on training/professional development of the project management community, where the PMO will:

- Identify competencies needed by high-performance Project Managers
- Provide Project Manager and team member training and development in professional methods, tools, and techniques to support career paths, as well as project leadership and team management
- Promote project management certifications
- Provide executive-level awareness of project management methodologies, and expert assistance in the form of coaching for functional staff involved in projects

5.7.4.3 The PMO Organization and Staffing – Sized to Type

The organization and responsibility of the PMO in the project management domain are to correspond to the type: basic, standard, advanced, or COPE.

A mix of skills and roles is needed to ensure that the PMO plays a central role in guiding the successful execution of strategic initiatives within the organization.

The PMO manager or director position, reporting to a steering committee sponsorship, is the most critical and should be equivalent to that of a high-level functional manager and is supported by professional and administrative personnel.

Structure and staffing, beyond the PMO manager, will vary depending on its type and will incorporate some or all of the following subject experts:

- Technical support staff developing and providing best practices, standards, methods, methodology and process development, and PM information systems
- Project estimator, project planner/scheduler, and resources administrator to secure resource skills to requirements, balance scarce resources, forecast, plan for resource shortfalls, and secure key resources
- Project risk management advisors
- Project controller for project oversight and review and quality assurance
- Qualified trainers for skills development
- Knowledge management coordinator for the collection and storage of project documentation and records, training, and lessons learned

- Project administrators for back-office tasks, report generation, and software support
- Procurement and contract administrator
- Financial analyst
- Other expertise adapted to the nature of programs/projects performed in the organization

At the basic PMO level, staffing will be restricted to the PMO manager and potentially one or two individuals. This will limit the scope of PMO service this type can provide.

PMO personnel headcount would increase in conjunction with the type's level of maturity, reaching a fully staffed state.

For COPE PMO organizational structures, which are the highest in rank, the PMO director manages and directs all corporate projects, including those that cross divisional boundaries, and is also responsible for resource distribution and funding allocation on all projects. At this stage, the PMO will possess a staff of Project Managers which will greatly increase program/project success.

5.7.5 Planning for the Introduction of a PMO

The formal implementation of a PMO is an important step for the organization. Similarly, the evolution from one level to the next has to be made equally in a formal manner.

This section will describe the scope of the planning and implementation of a newly created PMO and will concentrate on the steps to perform for its introduction to the organization. The initiated PMO will often be for a basic level, irrespective of the project management Maturity Level, as it is important that the organization adapts to this newly created functional entity. A series of initial steps have to be undertaken prior to the formal approval to initiate the PMO.

A major first step is to define the role, scope, and outcomes to be achieved by the PMO by consulting with major Stakeholders to collect their needs and to determine how and where the PMO will fit within the organization. Once this is established, initial funding would be required to conduct an As-Is assessment and review existing projects and skill levels to identify global needs and examine recent and existing projects to identify the state of methodologies and processes, PM competencies, and skills, as well as other gaps that the organization can address through the PMO. A To-Be requirements analysis is then conducted to determine the scope and contents of the PMO implementation project. A PMO Business Case, success factors, and performance measurement mechanism are agreed upon and a project plan, resource needs, and budget are established.

It is at this stage, with the collation of the above documentation, that formal approval can be given and a PMO manager assigned with an accompanying sponsor. As with all formal projects, a project charter is established defining the goals and

Objectives to be achieved and the Deliverables, resources, and funding, together with the scope of responsibility and authority for the PMO manager.

5.7.5.1 Establishing the PMO Road Map for Development and Implementation

The PMO project will establish a road map for the development and implementation of the approved scope contents. The road map will contain details on:

- Organization and staffing plan
- Infrastructure and facilities plan
- Methodologies and procedures development plan
- Competency plan for Project Managers and staff
- Implementation budget
- Implementation schedule
- Resource plan
- H/W and S/W acquisition development plan
- Consolidated PMO implementation plan
- Organization Change Management plan

The road map will also include the efforts required to develop processes specifically for use by the PMO to perform its own activities and conduct necessary training for the PMO staff (Figure 5.14).

For the project initiation and planning and roll-out, the PMO manager will require a limited staff, as well as a corresponding physical infrastructure.

The road map schedule highlights the major activities to perform over a previously agreed-upon timeframe. Care must be taken to avoid over-ambition. Experience has demonstrated that the launch of a PMO from its outset will have a duration of a minimum of 12 months, which can be extended up to 18 months. Individual PMO Deliverables may be produced during this timeframe; however, the complete set of those Deliverables, implemented in the project performance environment, will only occur at the end of the road map. Subsequently, the PMO will focus on enhancing and improving the support it provides to the project management community.

The road map will follow a standard life cycle approach, where PMO Deliverables will be made available to the organization according to a coherent integration of ongoing projects.

This will cover:

- Development and dissemination of processes, standards, and methodologies
- Providing project support expertise
- Project management competency and skill development

It is strongly suggested that the selected ongoing projects which may benefit from the early availability of developed processes, standards, and methodologies be

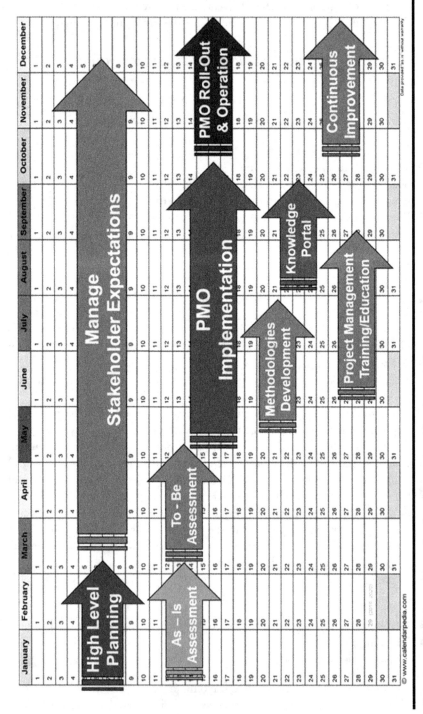

Figure 5.14 PMO road map for development and implementation.

identified previously. Similarly, specific training may be dispensed for selected projects to enhance their ongoing performance.

5.7.6 Implementation of the PMO

The PMO implementation is to be managed as a project, within the defined scope and building the activity schedule and resource needs. The major activities will be to integrate applicable organizational policies, establish technical processes/procedures, establish project management qualifications, and develop project classification guidance. A Change Management plan must be established for the deployment of the PMO Deliverables, and it must be shared with the project management community, sponsor(s), and major Stakeholders. Limitations will exist when insufficient PMO resources are available to perform these activities, and consequently, the timeframe for the implementation will be extended.

5.7.6.1 Developing and Deploying the PMO Processes and Procedures

This consists in constructing an initial inventory of all projects underway and gathering methodologies, processes, and procedures currently used by different projects. Special attention is to be given to the currently used:

- Project management processes, templates, and tools
- Charter, statement of work, or other concept description
- Benefits/cost model and financial analysis templates
- Scope management and associated change request management
- Requirements and specification documents
- Risk management forms, templates, and response techniques and plans
- Project planning and scheduling templates and supporting S/W tools
- Test plans, deployment plans, and support plans
- Standardized status reporting formats
- Post-project audit programs

From the current information gathered from ongoing projects and previously documented projects, collected methodologies, processes, templates, and other project-related documents are analyzed. The PMO must then conduct a series of workshops with Project Managers to gather their input as to which processes are relevant and where improvements or abandonments are required and agree on how to proceed. The work then proceeds to develop the set of project management methodologies pertinent to the organization and the ongoing and future projects. A repository library of PMO materials is constituted which will contain the agreed-upon project management methodologies, processes, and templates.

Special attention will be given to standard progress report formats, processes for gathering and reporting the estimates, and actual financial results of projects. Project performance dashboards may also need to be developed, in agreement with the PMO sponsor and upper management.

Deploying the PMO methodologies and processes will follow the agreed-upon sequence as discussed and agreed with the Project Managers during the initial workshops. Certain projects would be selected to be the "test-beds" of the new or adapted methodologies and processes. Complementary training on these processes will be dispensed when necessary. Feedback and modifications from the use of the processes can then be made, and the updated versions can be progressively applied to other ongoing projects. The lessons learned will be collected and stored in the repository library.

Note that the initial deployment schedule may be adjusted to accommodate results and evolutions from ongoing projects, as well as changes in the organizational environment that may require specific attention.

The PMO manager will report to the PMO sponsor and major stakeholder on the deployment progress, according to a previously agreed-upon schedule frequency, including dashboards for senior management and functional managers, showing the status of all active projects. All project reports are to be shared with Project Managers.

5.7.6.2 Conducting the Project Manager Competency Model Assessment

A project management competency assessment is to be conducted to ascertain the knowledge gaps to be filled for the project management community – Project Managers, project team members, and functional staff solicited to perform on projects.

This assessment identifies the level of competency/skill of a Project Manager in the following categories: technical, personal, and business/leadership, and will focus on the Project Manager's current competency, skill, and experience in the following areas:

- Organizational understanding
- General project management aptitude
- Project management knowledge areas (with reference to PMI)
- Personal communication, leadership, and motivational skills

Scoring will use the table shown in Figure 5.15.

And an example of the questions is shown in Figure 5.16.

The results will indicate the areas that need to be addressed through education and training.

Scoring:	1	2	3	4	5
Ability	None	Low	Conversant	Proficient	Expert
Knowledge	No	Some	Average	Knowledgeable	Expert
Use	Not at all	Sporadically	Occasionally	Frequently	Always

Figure 5.15   Project Manager Competency assessment – scoring.

	Ability to	1	2	3	4	5
1.	Ability to implement and apply project management processes, best practices, and techniques within the framework of the organisation's methodology(ies) throughout the project life cycle					
2.	Conduct ongoing analysis to identify and forecast budget and schedule variances;					
3.	Develop, and maintain, a formal and comprehensive project plan that integrates and documents the project work: typically includes the work breakdown structure, project deliverables/milestones, acceptance criteria, schedule, budget, communication, risk, and procurement plans					

Figure 5.16   Project Manager Competency model assessment – sample questions.

The PMO will initially collect and collate distributed assessments from Project Managers and plot the Maturity Level of management of projects expertise. This will enable the PMO, in coordination with Project Managers, to establish a gap analysis for the Project Manager pool, identifying the areas to strengthen and developing a general education plan to cover the total audience. Individual training/education plans are discussed and developed with each Project Manager, and certification needs are identified and planned for the learning phase and subsequent examination.

5.7.6.3 Performing PM Training and Education Plan: Project Managers

The PMO will develop a general project management curriculum and establish a training calendar aligned with the availability of Project Managers. Funding must be secured in collaboration with the HR function. In conjunction with the line managers to whom the Project Managers report, the proposed training curriculum and schedule are aligned with the individual's career progression plan. A plan for certification exam preparation is established with each Project Manager and agreed upon with the line manager.

The planned training can then be managed by the PMO staff and conducted internally if qualified instructors are part of the PMO team. Else, qualified external providers are secured to perform the training courses

5.7.6.4 Performing PM Training and Education Plan: PMO Staff and Non-Project Staff

To ensure cross-organizational knowledge in project management, the PMO will also develop a general curriculum for PMO staff and establish a training calendar for staff training, delivered internally or with external providers.

For non-project staff, the PMO will establish a curriculum of introductory courses on project management and present this curriculum to management and functional managers for approval and funding.

5.7.6.5 Determining Extent of Project Portfolio Management

During the "To-Be" assessment, the PMO will define the extent of its responsibility for Project Portfolio Management. The overall responsibility and authority of the PPM will greatly depend on the Maturity Level of the organization and the type of PMO it currently operates. Please see Section 5.7.2.4 for a detailed discussion on this topic.

As the PMO has the responsibility to develop and deploy methodologies for the management of projects, and Project Portfolio Management is included in this area of responsibility, it will develop and deploy, along with the PPM team, the necessary processes. These have already been identified above in Section 5.4.

5.7.6.6 PMO Implementation Issues to Address and Overcome

The introduction of a PMO will face many challenges. Management enthusiasm will dwindle if the PMO is perceived as not being effective and its value is questioned; additionally, the culture present in the organization and project management community may resist change. The PMO manager must ensure that the focus is directed at the necessity of the organization to establish professionalism in the management of projects and demonstrate the importance of the PMO to do things right. The focus is multi-pronged, concentrating on the transparency of the projects in the portfolio and high-quality project execution performance, including coherent funding and resource allocations. This focus must be shared with senior management, functional management, and the Project Managers and project teams.

  I. Focus on project transparency
   - Aligning the Right Projects to the Organizational Strategy
   - Employing effective benefit realization metrics on projects
   - Utilizing robust project selection and governance processes
   - Use of reliable and structured project processes
  II. Focus on high-quality project execution performance
   - Enhancing competence and skills of Project Managers and team members
   - Deploying effective methodologies and processes
   - Provision of early warning signals from project review processes

- Assessing post-project performance
- Institutionalizing project lessons learned

III. Focus on the relationship between PMO and the Project Managers

This is of major importance to avoid challenges as to the role and effectiveness of the PMO. Project Managers will raise a variety of questions, especially the seasoned ones, covering but not limited to:

- Does the PMO support all business strategic initiatives?
- Is the PMO considered effective by the project management community?
- Is the PMO a strategic facilitator and integrator?
- Does the PMO engage in portfolio management activities?
- Does the PMO leadership have direct access to and guidance from top decision-makers?
- Are projects selected objectively?
- Is project evaluation or audit conducted effectively?
- Is project financial performance measured?
- Are statistics or scorecards about projects' success maintained?

It is essential that the PMO manager understand the scope of responsibility and is at comfort with responding to the above questions.

IV. Facilitating stakeholder buy-in for the PMO

The PMO manager must constantly promote the essence and effectiveness of the PMO. Senior-level management commitment is very crucial to set up and "incubate" the PMO in the initial phase. As the expectations of the PMO evolve as time passes by, disenchantments can occur.

The PMO manager must articulate and demonstrate the value of professional Management By Projects and emphasize the benefits gained by formal processes. The target is wide and covers senior managers and functional managers as Stakeholders. Scheduled invitations to PMO forums are of great value to present the PMO's role and value and to discuss Stakeholders' issues.

Transparency in *what* the PMO brings to the organization, as functions and benefits, has to be clearly articulated by presenting contents and context to:

- Standardization of project management methodologies, tools, and templates
- Maintenance of a best practices library
- Deploying processes for selecting Project Managers and teams
- Availability of a training curriculum for Project Managers
- Existence of a coherent resource management process for assignments of project teams, with participants from multiple business functions and disciplines
- Management of the organization's project portfolio by the PMO or by the PPM
- Use of a centralized project reporting within the PMO, depending on the type of PMO
- Service offered by the PMO as a pivot for communicating with all internal and external Stakeholders

V. Focus on the PMO's key performance indicators

The PMO manager must seek to quantify *how* the PMO contributes effectively to projects by presenting actual data as to how projects are performing to a higher standard. Tangible and Intangible Benefits are to be underscored by reporting on and illustrating the effectiveness of projects in:

■ Realization of Business Benefits
■ Coherent governance of projects
■ Performing the Right Projects for the right purpose
■ Centralized project alignment
■ Competent and skilled Project Managers
■ Standardization of processes leading to high efficiency
■ Lower budget spends
■ Enhanced accuracy in delivery
■ Efficient resource usage
■ Comprehensive reporting
■ Shared lessons learned

5.7.6.7 PMO Operational Costs

PMO operational cost must be funded and the approval for its yearly budget must be given at its first introduction. The yearly funding allocation is reviewed in a similar manner to a shared service department. Depending on the type of PMO operated, operational costs will cover:

Staff	Facility	Materials
Project Manager	Office	Training
Specialists/admin	Data/storage	Promotional
Trainers	Training room	Process documentation
Help desk	Computer network	
Tools developers	Software	

Certain organizations may opt to operate the PMO like a "business", generating demand through effectiveness and being funded by:

■ Global projects – by the global capital budget allocation
■ Internal projects – by department sponsoring the project
■ Customer projects – by the solution-selling entity

5.7.6.8 The PMO and Project Portfolio Management

Please refer to Section 5.9, "Applying a Combined PPM/PMO Structure", for an extended discussion on this topic.

From the To-Be A assessment, the following would be determined:

- Portfolio management process to develop and employ
- Accountability and responsibility assignments for programs/projects managed within the PMO portfolio
- Interfaces required for existing Project Portfolio Management systems
- Program/project set to be managed in a portfolio by the PMO

The PPM scope is to confirm business strategy, drivers, and business value, and utilizes a portfolio project selection criterion. This will be assisted by the PMO that will have the responsibility of developing a project classification guidance for programs/projects in the portfolio and establishing prioritization norms and assessment tools covering:

- Economic and financials
- Balanced scorecard
- Complexity
- Risk
- Schedule
- Resource demand
- Security
- Other as determined by:
    - Business perspective
    - Financial perspective
    - Performance perspective

5.7.6.9 Multiple PMOs Across the Organization

To address both transformational and tactical change program/project needs, organizations may seek to institute multi-tiered PMOs to support the various business units and investments. Choices will depend on many factors; however, the most prominent will be the size of the organization, the number of ongoing and future programs/projects they manage, the Maturity Level of the enterprise, and the volume of CAPEX and OPEX involved.

Certain enterprises may opt to institute a single PMO that operates as a unique entity within the organization, while other enterprises will establish multiple PMOs that are operating independently at the strategic and tactical levels, be organizationally aligned, and/or are based on the division of PMO functional responsibilities by functional entity.

Senior management may decide to institute a corporate or enterprise PMO, overseeing the multiple PMOs, to ensure that programs/projects across the organization are supported in a similar fashion.

When multiple PMOs exist, they need to share common best practices, while allowing for individual PMO structures that serve best the interests of their functional group.

5.8 Applying a Combined PPM/PMO Structure

Enterprise strategy at the executive or operational level is demand management, where managers decide on which initiative to pursue and then launch the corresponding program/project (Figure 5.17).

When deploying a Project Portfolio Management structure, the organization needs to assess its level of maturity, as discussed earlier in this chapter. This will allow us to determine the scope and responsibility to be covered by the PPM. Additionally, and according to the established Maturity Level, the type of Project Management Office will be decided, ranging from basic to Center Of Project Excellence.

When it is considered that the PMO is still in its early stages of development, the organization may opt for a structure placing the PMO as a support function to the PPM, as illustrated in Figure 5.18.

As the organization's Maturity Level increases and value is demonstrated in both the PPM and the PMO, then the organization can be deemed to have reached the level of a COPE and be structured as illustrated in Figure 5.19.

A combined Project Portfolio Management and Project Management Office will comprise the disciplines of strategic alignment, prioritization, and governance of initiatives, programs, and projects. Both will combine to answer the key strategic questions: "Are we Doing the Right Things?", "Are we doing the things right?", and "Are we realizing the benefits?".

Figure 5.17   Strategy demand framework.

Figure 5.18   Project Portfolio Management supported by the PMO.

Figure 5.19   The COPE (Center Of Project Excellence).

The PMO COPE will contribute and provide a structure to:

■ Ensure business unit alignment with short- and long-term business plans and strategy
■ Establish an enterprise-wide process to ensure the business units are involved in the rationalization and prioritization of investments
■ Ensure only investments with the greatest business value are funded
■ Identify opportunities to leverage investments across the organization
■ Provide timely approval of project investments

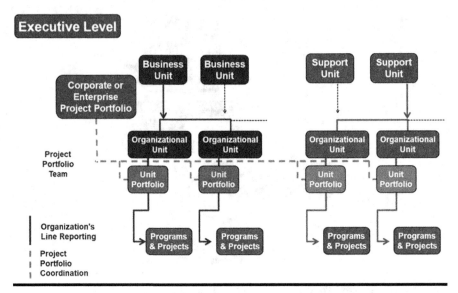

Figure 5.20   Multiple Project Portfolio Management entities in the organization.

5.8.1 *Multiple Project Portfolios in the Organization*

An organization will establish multiple PPMs to oversee transformational changes at the executive level and tactical changes at the operational level, be it a business unit or an organizational unit. The importance is that they will all function with the same enterprise portfolio core principles, while the operational levels will apply specificities to certain processes it will perform.

Thus, the PPM is a multitier system consisting of strategic, functional, and operational organizational portfolio components and is planned and executed at all these tiers in the organization (Figure 5.20).

5.9 Challenges for Management By Projects by PPM and PMO

5.9.1 *Challenges for the Project Portfolio Management – PPM*

The focus of Project Portfolio Management is to ensure programs/projects achieve the organization's Benefits Realization. PPM fulfills and manages stakeholder expectations by aligning programs/projects to the changing realities forced upon the organization by political, economic, social, technological, and environmental factors. PPM adds value to the organization as programs/projects are launched based on robust Business Cases and ensures organization-wide communication on programs/projects.

A coherent governance must be established that encompasses the structures, accountabilities, policies, standards, processes, and control mechanisms that are the basis for decision-making for initiatives and programs/projects. This will provide a decision-making framework that is logical, solid, and repeatable to govern an organization's funding and investments, where strategic and tactical goals remain the guiding lines.

As a challenge, Project Portfolio Management governance is to overcome the deficiency in the required competencies for managing project portfolios and the lack of mandate and decision-making power. A selection of these challenges requiring strong PPM focus are:

- Ensuring Management of Programs/projects competence
- Assigning appropriate decision-making power
- Establishing mandate and ownership of transformational and tactical changes
- Collecting and aggregating pertinent data
- Overcoming internal politics and power struggles
- Operating with clear processes and procedures
- Deploying guidelines and methods/models
- Avoiding too many interdependencies between programs/projects
- Preventing too many programs and projects

5.9.2 PPM – Business Alignment and Business Benefits

The project portfolio will set priorities, investment decisions, and resource allocation.

The alignment of projects with the overall strategy is continuously reviewed as projects are quantified by value to the organization, business drivers, measurable Deliverables, and ranked and prioritized.

Continuous improvement projects are launched to enhance operational methodologies and processes and are strategic when they deliver against organizational business needs. These will focus on linear, incremental improvements within an existing process.

Project Portfolio Management can effectively support continuous improvement projects as strategic enablers of the business. Additionally, the Project Portfolio Management will establish and conduct its own improvement plan:

- Defining the Objectives expected to be achieved from portfolio management
- Guiding, measuring, and prioritizing improvement activity
- Allowing predefined metrics to drive results and accountability for performance improvement

5.9.3 PPM – Identification and Involvement of Key Stakeholders

Executives and line managers have positive/negative financial or emotional interest in the outcome of their organizational performance. They have key motivational drivers and their professional and personal interests will influence their decisions, as their interests will often go beyond that of the program/project.

It is therefore important that the PPM facilitates stakeholder buy-in by assessing and establishing with different Stakeholders the appropriate type of participation required at the successive stages of the Project Portfolio Management cycle. The communication will underline:

- The problems that the program/project is seeking to address (at the identification stage)
- The purpose of the program/project (once it has started)
- The exchanges with the Stakeholders on the project portfolio's value to their interests and the program/project
  - Meeting program/project expectations
  - Benefit assurance from successful program/project completion
  - Regular reporting to ensure alignment

5.9.4 Challenges for the Project Management Office – PMO

The challenges to operating any type of PMO are varied. The primary underlying factor is the organization's Maturity Level in project management across the enterprise. Ambitious launches of a PMO incompatible with the state of maturity will not only be ineffective but will also taint its principal essence.

As the PMO may well be seen as an overhead and be difficult to measure as to the value it brings to the organization, it is imperative that the PMO manager and PMO sponsor be the promoting voices of the structure to management at all levels and program/project Stakeholders, including the management of projects community.

The PMO must be staffed appropriately to provide the project management methodologies and training to the project management community and offer the subject-matter expertise required. A lack of staffing will only result in a deficiency in the PMO's functions and create objections and resistance to its continuing institution.

5.9.5 Project Management Office Metrics and Reporting on Projects

The premise is that management understands project management basics. Metrics and indicators are to be provided on the performance of the PMO and must show that projects are well under control. The PMO must demonstrate that its function assists in the creation of business value, expressed in cost reductions or revenue increases for the organization, as facilitated by successful project deliveries enabling the performing operational functional units to meet their Objectives and Business Benefits. The functional performance of the PMO must demonstrate the quality of the PMO functions, improvements in its operation, and the quantitatively measured enhancements introduced within the projects using PMO-generated methodologies and processes.

Reporting on ongoing and completed projects will consolidate metrics collected to demonstrate that:

- Established project scope is agreed upon between the organization and project performance groups
- Projects are managed with up-to-date and realistic time, cost, and scope projections
- Project schedules have accurate estimates that are not constantly revised
- A decision-making process exists for addressing and resolving project conflicts

5.9.6 PMO: Continuous Improvement

The PMO will seek to ensure that the organization effectively functions in terms of the management of projects and determine if the organization is ready to move to the next project management Maturity Level.

The PMO will need to focus on enhancing and improving:

- Value realization through strategic alignment of projects
- Customer satisfaction through superior PMO performance
- Process excellence through compliance with project management policies and procedures

Following the deployment and use of methodologies and processes, continuous improvement is to be performed on a regular basis according to the results collected from ongoing programs/projects. Corrective steps are identified and discussed with Project Managers as to the most appropriate manner in which to incorporate changes that will bring a greater impact on the delivery of the program/project. An action plan is to be established, identifying a list of corrective actions, and presented to the PMO sponsor and steering committees for approval. Additional funding may then be required to complete the continuous improvement actions.

Index

Pages in *italics* refer to figures and pages in **bold** refer to tables.

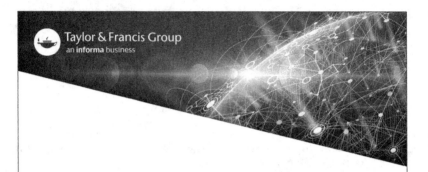

Printed in the United States
by Baker & Taylor Publisher Services